石油石化企业现场安全督导系列丛书

危险有害物质及火灾爆炸保护

Hazardous Substances and Fire & Explosion Protection

赵宏展　贺晓珍　杨意峰　编著

中国石化出版社

内 容 提 要

本书为《石油石化企业现场安全督导系列丛书》之一，介绍了危险有害物质以及火灾爆炸保护共计两组知识模块，包含化学品安全、硫化氢(有毒气体)安全、建筑物火灾保护等5个专题，可作为做好管理人员现场安全督导需要了解的支持性信息。

本书内容为中英文对照，便于读者将中文和英文结合在一起阅读，以便准确理解其内容。

本书可用于石油石化企业各级管理人员自学或培训，也可作为其他行业的管理人员以及石油石化大专院校工科专业及其他院校安全相关专业师生的安全管理类参考书。

图书在版编目（CIP）数据

危险有害物质及火灾爆炸保护／赵宏展，贺晓珍，杨意峰编著．—北京：中国石化出版社，2019.8
（石油石化企业现场安全督导系列丛书）
ISBN 978-7-5114-5449-2

Ⅰ.①危… Ⅱ.①赵… ②贺… ③杨… Ⅲ.①石油化工-有害物质-防护②石油化工-防火③石油化工-防爆 Ⅳ.①TE687

中国版本图书馆CIP数据核字（2019）第170230号

中国石化出版社出版发行

地址：北京市东城区安定门外大街58号
邮编：100011　电话：(010)57512500
发行部电话：(010)57512575
http://www.sinopec-press.com
E-mail:press@sinopec.com
北京柏力行彩印有限公司印刷
全国各地新华书店经销

*

850×1168毫米32开本4.625印张104千字
2019年11月第1版　2019年11月第1次印刷
定价：26.00元

前言
Preface

企业健康安全环境(HSE)管理工作的属地管理和直线责任原则已经深入人心，很多企业也高度重视“有感领导”(指有安全认知的领导)，不断强化各级管理人员对HSE工作的示范作用。越来越多企业的管理人员经常深入生产作业现场，实地调研和指导HSE管理工作，践行“有感领导”。随着企业HSE管理体系越来越成熟，不同角度、不同深度、不同方式的各级管理人员开展的HSE检查和HSE审核也越来越丰富多样，对于生产作业场所的HSE风险管控起到了至关重要的作用。

上述背景产生了两个问题：一是石油和石化企业各级管理人员在企业HSE管理体系框架下如何更精准有效地发力？二是石油和石化企业有哪些共性的HSE知识模块值得总结和提炼？

编著者基于10余年的海外项目工作经验，在广泛调研国内外大量文献资料的基础上，通过跟诸多资深项目管理人员和资深HSE专业人员的沟通交流，编写了本丛书。本丛书呈上了对上述两个问题的初步回应，推荐了基于“安全交谈(Safety Conversation)”的“管理人员现场安全督导方案(Leadership Site Visit Program)”，包括：具体做法、注意事项、支持性信息(知识模块)。

在安全氛围(Safety Climate)的三"心"模型[1]基础上，我们坚信：基于"安全交谈"的管理人员现场安全督导，可以通过管理人员对于安全问题的"关心"，激励员工对于安全问题的"上心"，并最终营造企业或项目对于安全问题的"上下一心"。

人们期待"安全保障"，并提出了"安全保障"的需求和要求。实际上，只有"共建安全"才能获得更加牢靠的"安全保障"，因为"安全保障"的需求者和供给者是不可分离的。在属地管理、直线责任、"有感领导"的基础上，人们还必须坚持"安全共建"和"安全共享"的原则，即每个人都坚信"我需要安全，安全需要我"。

本丛书呈上的中英文对照的支持性信息(知识模块)共分为八组，分别是：

(1) HSE 风险管理工具和过程；

(2) HSE 风险贡献因素和升级因素；

(3) 旅程管理和道路交通安全；

(4) 人员健康和保护；

(5) 工作场所常见危害；

(6) 潜在高后果活动；

(7) 危险有害物质；

(8) 火灾爆炸保护。

---

[1]于广涛，李永娟．安全氛围三"心"模型的构建与检验[J]．中国安全科学学报，2009，19(9)：28.

本丛书“臻选”的支持性信息(知识模块)融入了以下内容：(危害)控制层序(Hierarchy of Controls)的原则、设计保障安全(Prevention through Design)的做法、领结图(Bowtie Diagram)的精髓、保护层分析法(Layer of Protection Analysis)的思路、屏障(Barrier)的要义。此外，安全例证(Safety Case)不仅仅是英国等发达国家法律规定的具体做法，更是一种风险管理思路。危险轨迹(Line of Fire)既是生产作业场所的一种“常态”，也是很多人身伤害的本质，目前已经作为国际油气生产商协会(IOGP)在2018年发布的九条保命法则(Life-Saving Rules)之一而更加受到重视。管理巡查(Management Walkthrough)、管理层检查(Management Inspection)和管理审核(Management Audit)，三者密切相关，也经常被混淆，本丛书对其区别和联系也做出了解读。

管理人员现场安全督导推荐做法、常用安全模型以及八组支持性信息(知识模块)等内容顺序编排在本丛书的相应分册当中。四个分册的名称具体如下：

- 《安全督导推荐做法及常用安全模型》
- 《基本HSE风险因素及管理》
- 《工作场所常见危害及潜在高后果活动》
- 《危险有害物质及火灾爆炸保护》

本丛书的英文内容参考了以下政府组织和行业协会公开出版的英文文献资料，在此表示诚挚感谢！

◇ 英国健康安全执行局(UK HSE, UK Health and Safety Executive)

◇ (美国)职业安全与卫生管理局(OSHA, Occupational Safety and Health Administration)

◇ 美国化学工程师协会(AIChE, American Institute of Chemical Engineers)

◇ 国际油气生产商协会(IOGP, International Association of Oil & Gas Producers)

◇ 国际钻井承包商协会(IADC, International Association of Drilling Contractors)

本丛书的写作还参考了其他大量国内外文献资料,在此对原著者也深表感谢!

回望过去,本丛书英文内容的最终定稿是一个漫长的积淀和筛选过程,历时10余年时间。2010年1月,编著者之一刚刚取得安全工程博士学位,作为唯一的中方人员,有幸被指派组建一支国际化的HSE团队,服务于一个员工来自30多个国家的中东地区油田开发生产项目。先后有来自英国、美国、法国、加拿大、澳大利亚、新西兰等国家的60多位资深HSE专业人士以油田员工或知名咨询公司顾问的身份,成为这支国际化HSE团队的一员。2012年12月和2013年6月,本丛书的另两位编著者先后加入这支HSE团队,成为协调和领导这支国际化HSE团队的核心力量,但英文仍然是覆盖整个HSE团队的唯一工作语言。在开放、包容、交流、共享、相互尊重的基础上，这支 HSE 团队整体所能提供的 HSE 支持和服务实现了最大程度的国际接轨。总之，本丛书可以说是编著者经过 10 余年开放融通的海外项目磨砺后交上的一份答卷。

本丛书“臻选”的支持性信息(知识模块)采用中英文对照的形式编写和排版，目的是希望读者能够将中文和英文结合在一起阅读，提取其中的“最大公约数”，从而更加精准地理解本丛书所提供信息和知识的本意，更好地辅助和支持管理人员开展现场安全督导。

本丛书最终得以公开出版得到了众多资深管理人员、中外专家和 HSE 同仁的支持和参与，凝聚了众人的智慧和心血，编著者不过是在知识的沙滩上捡拾贝壳的三人，我们虔诚地把认为有价值的知识贝壳收集起来，经过筛选和加工以中英文双语的形式呈现给广大读者。感谢默默无闻支持编著者完成本丛书的所有人！

本丛书主要适用于石油石化企业各级管理人员自学或培训之用，也可作为其他行业的管理人员以及石油石化院校工科专业及其他院校安全相关专业师生的安全管理类参考书。

管理人员现场安全督导工作有着很强的知识性和实践性，书中难免存在疏漏之处，敬请批评指正。如有任何意见和建议，欢迎联系以下电子邮件地址：safetyview@ 163. com。

目录

附录

Contents

1

HAZARDOUS SUBSTANCES

危险有害物质

1.1 Chemicals Safety

1.1.1 Prompts

(1) Have all chemical hazards been identified, risks assessed and used to establish safe methods of work?

(2) Is a chemicals management procedure in place and operational?

(3) Are any chemicals inappropriately stored in unmarked containers or unsuitable containers?

(4) Are spark-free storage spaces provided for flammable substances away from ignition sources?

(5) Is proper storage practised to minimize risk of fire including spontaneous combustion of hazardous materials and waste-including exposure to sunlight?

(6) Are all outdoor combustible materials stored more than 3m from a building or structure?

(7) Is the Safety Data Sheet for Chemical Products (SDS) in place and operational?

(8) Are chemicals clearly labelled in line with their hazardous properties using an internationally recognised labelling system?

(9) Are all hazardous materials in storage legibly labelled and protected from spillage or leakage?

(10) Do site staff understand the firefighting requirements of chemicals?

(11) Do site staff understand which chemicals need to be cooled during a fire to prevent escalation?

1.1 化学品安全

1.1.1 督导提示

(1) 是否识别出了所有化学品危害？是否对其进行了风险评估并据此确立了安全的工作方法？

(2) 是否建立了化学品管理程序并有效实施？

(3) 是否有化学品被不适当地储存在没有标签或者不合适的容器中？

(4) 是否为易燃物质提供了远离点火源的无火花储存空间？

(5) 是否采用了适当的储存(规范)以最小化火灾风险？包括(最小化)危险物料和废弃物的自燃(含暴露于阳光下的情形)。

(6) 所有户外可燃物料是否存储在建筑物或构筑物的 3m 以外？

(7) 化学品安全技术说明书(SDS)是否到位并有效？

(8) 是否采用国际认可的标签制度并根据化学品的危害特性对化学品进行清楚地标注？

(9) 存储的所有危险物料是否进行了清晰标注并加以保护以免溢出或泄漏？

(10) 现场人员是否了解化学品的消防要求？

(11) 现场人员是否了解哪种化学品在着火时需要冷却以防止后果升级？

(12) Are fire extinguishers and spill kits used at chemical storage areas appropriate, functional and well maintained?

(13) Are spills of flammable or combustible materials cleaned up promptly?

(14) Do site staff understand the first aid requirements for each chemical?

(15) Have the PPE requirements for all chemicals been defined?

(16) Is the necessary chemical PPE available on site?

(17) Is a process to assess new chemicals in place and operational?

(18) Is a process for tracking and safely disposing of expired chemicals in place and operational?

(19) Is there a process in place and operational to ensure that personnel who work with chemicals are competent?

(20) Is a process to monitor, audit and improve the performance of the Chemicals Management System in place and operational?

1.1.2 Reading

Chemicals underpin vital aspects of asset performance, including process safety, asset integrity and environmental performance. Some chemicals are intrinsically hazardous to both people and the environment. Chemical hazards arise in the form of solids, liquids, gases, mists, dusts, fumes, and vapours that exert harmful effects by inhalation (breathing), absorption (through direct contact with the skin or through wounds), or ingestion (eating or drinking).

(12) 化学品存储区的灭火器和泄漏清理装备是否合适？功能是否正常并维护良好？

(13) 易燃或可燃物料的溢撒是否得到立即清理？

(14) 现场人员是否了解每种化学品的急救要求？

(15) 是否确定了所有化学品对个体防护装备(PPE)的要求？

(16) 必要的针对化学品的个体防护装备(PPE)在现场是否可以获取？

(17) 新化学品评估程序是否建立并有效实施？

(18) 化学品追踪和过期化学品安全废弃程序是否建立并有效实施？

(19) 确保化学品工作人员胜任的程序是否建立并有效实施？

(20) 监测、审核和改进化学品管理制度实施效果的程序是否建立并有效实施？

1.1.2 知识准备

化学品支撑设施性能表现至关重要的一些方面，包括过程安全、设施完整性和环境绩效。一些化学品从本质上对人和环境具有危害性。化学品危害以固体、液体、气体、薄雾、粉尘、烟气和蒸气的形式出现，通过吸入(呼吸)、吸收(直接接触皮肤或通过伤口)或食入(饮食或饮水)(对人体健康)施加有害影响。

Hazards associated with chemicals should be mitigated by effective chemicals management that ensures location specific HSE controls are in place and are adequate.

- **A process should be in place to identify chemical hazards, assess risks and to establish a safe method of work**

HSE risks associated with each stage of the chemical's life cycle (identify the need for a chemical, identify the range of chemicals to meet the need, select the preferred chemical, obtain the chemical, transport the chemical, store the chemical, use the chemical and safely dispose/reuse the chemical) should be identified and managed. Each activity of each stage of the chemical's life cycle should have control measures in place covering both prevention of exposure and emergency response in case of exposure.

HSE risk assessments should be performed for all chemicals used in site work activities. Competent HSE advisors should be responsible for performing the risk assessments which should be supported by supervisors and workers.

All work involving chemicals should be avoided if reasonably practicable. If unavoidable, the risk assessment should consider all measures that can be implemented to reduce the risk to a tolerable level.

The chemical risk assessment should consider the likelihood of exposure, the exposure pathway and the severity of the consequence of exposure. The severity is based on the toxicity of the chemical and the dose which consists of the amount/concentration of chemical and duration of time exposed to the chemical.

Control measures should be based on the hierarchy of controls:

① Eliminate the use of a harmful substance completely.

应通过有效的化学品管理，确保针对具体工作场所的化学品 HSE 控制措施到位且足够，减轻化学品相关的危害。

• 应建立程序辨识化学品危害，评估其风险并确立安全使用化学品的方法

与化学品生命周期(包括识别化学品需求；识别满足上述需求的化学品的范围；择优选择化学品；获取、运输、储存、使用和安全处置/再利用化学品)每个阶段相关的 HSE 风险都需要加以辨识并进行管理。化学品生命周期每个阶段的每一项活动的控制措施都应到位，包括预防(人员和环境)暴露(于化学品危害)以及一旦暴露后的应急响应。

现场工作活动使用到的所有化学品，均应开展 HSE 风险评估。胜任的 HSE 人员应负责在主管人员和工作人员的支持下开展风险评估。

如合理可行，应避免所有涉及化学品的工作。如不能避免，风险评估应考虑可实施的所有措施，以把风险降低至可容许的水平。

化学品风险评估应考虑发生暴露的可能性、暴露途径和暴露后果的严重性。该严重性是基于化学品的毒性和剂量，剂量又包括化学品的数量/浓度以及暴露时间。

控制措施应基于(危害)控制的优先层序：

① 彻底消除有害物质的使用。

② Substitute the harmful substance with a safer one.

③ Use a safer form of the substance, e. g. paste rather than powder.

④ Change the process to release less of the substance.

⑤ Enclose the process so that the substance does not escape and affect workers.

⑥ Extract emissions of the substance near the source.

⑦ Have as few workers in harm's way as possible.

⑧ Provide personal protective equipment (PPE) such as gloves, coveralls and a respirator that fits the wearer.

⑨ Provide emergency and spill response equipment.

Control measures are always a combination of equipment and working practices to reduce exposure. The right combination is crucial. No measures, not even practical ones, can work unless they are properly implemented and are used consistently and correctly by the workers.

Safety Data Sheets for Chemical Products (SDSs) should be used to inform the risk assessment. Emergency and spill response information provided on the SDSs should be used as the basis to identify emergency and spill response measures. SDSs should be displayed in all areas where chemicals are stored or used. Copies of the SDSs of all chemicals used should be available at all the medical facilities.

Chemicals should be approved for procurement and use both technically and from an HSE risk perspective by a competent technical authority or delegated expert.

②用更安全的物质替代有害物质。

③ 使用有害物质更安全的形态，如使用膏剂代替粉剂。

④ 改变工艺流程以减少有害物质的释放。

⑤ 封闭工艺流程以防止有害物质漏出并影响工作人员。

⑥ 在有害物质源头附近收集其排放物。

⑦ 尽可能减少接触有害物质的工作人员。

⑧ 提供个体防护装备(PPE)，如手套、连体防护服和适合使用者佩戴的呼吸器。

⑨ 提供应急和泄漏响应装备。

控制措施常常是用来减少暴露的设备和工作方式的组合。正确组合是至关重要的。即使操作性很强的控制措施，如果没有被工作人员恰当执行并持续正确应用，也无法发挥它们的作用。

化学品安全技术说明书(SDS)应被用来为风险评估提供信息。应以化学品安全技术说明书(SDS)提供的应急和泄漏响应信息为基础，识别应急和泄漏响应措施。化学品安全技术说明书(SDS)应张贴在化学品储存和使用场所。现场所有医疗设施内也应该有所有化学品的化学品安全技术说明书(SDS)备份。

化学品的采购和使用应从技术和 HSE 风险方面获得胜任的技术负责人或受委托专家的批准。

• A chemicals management procedure should be in place and operational

A fully documented Chemicals Management Procedure should be available and fully integrated into the company's HSE Management System. The Chemicals Management Procedure should be risk-based and covers each stage of the chemical's life cycle. Site should be aware of the procedure and be able to demonstrate how it is being used.

Site should ensure that all chemicals have been risk assessed and that appropriate storage and handling requirements (i. e. usage procedures) are in place. Incompatible chemicals should not be stored together.

All chemicals should be clearly labelled in line with their hazardous properties using an internationally recognised labelling system.

All chemicals should have supporting SDSs displayed at the storage and work areas which should be understood by workers who use or handle the chemical. SDSs should also be available to all workers in the event of a spillage or emergency situation.

Chemical control measures must be inspected and maintained to ensure they work as planned. In particular, the site should ensure that the:

① Chemicals in storage or chemicals used in the process aren't emitting uncontrolled substances.

② Chemical control equipment continues to work as it was designed.

③ Workers comply with chemical usage procedures.

- **应建立化学品管理程序并有效实施**

充分文件化的化学品管理程序应可获取，并充分整合到企业[1] HSE 管理体系中。化学品管理程序应基于风险并涵盖化学品生命周期的每个阶段。工作现场应了解并能展示如何使用该程序。

工作现场应确保所有化学品都进行过风险评估，并确保恰当的储存和使用要求(如：使用程序)到位。不相容的化学品不可以共同存放。

所有化学品应采用国际认可的标签制度并根据它们的危害特性进行清楚地标注。

所有化学品都应有配套的化学品安全技术说明书(SDS)并张贴在化学品储存和使用场所，使用或处理化学品的工作人员应理解相关信息。在发生化学品泄漏或紧急情况下，所有工作人员都可以获得化学品安全技术说明书(SDS)的信息。

化学品控制措施必须得到检查和维护，以确保其按计划执行。工作场所尤其应确保：

① 存储的化学品和工艺过程使用的化学品不排放非受控物质。

② 化学品控制设备按设计要求正常工作。

③ 工作人员遵守化学品使用程序。

---

[1] compang 也可称为公司。

PPE is often used as part of control measures. PPE should be periodically checked and maintained. Failed PPE that no longer provides adequate protection against the chemical hazard should be removed from service for repair or disposal.

Sites should ensure that appropriate fire extinguishers and spill kits are available in chemical storage areas and that appropriate signage is used.

Chemical usage should be tracked from procurement to expiry. Expired chemicals should be identified and disposed of in a safe manner in line with company Waste Management procedures. This should include disposal of packaging materials.

- **A process should be in place to ensure that personnel who work with chemicals are competent**

Site should be able to demonstrate a process is in place to ensure that whoever designs, installs, maintains and tests chemical control measures have the necessary skills, knowledge and experience and are competent.

Workers should be fully involved in the development of chemical control measures to make sure they are suitably equipped to carry out the work. Workers should be encouraged to suggest improvements, and to report anything they think might be going wrong.

Workers, and anyone else who needs to know, should be provided information and full instruction on the chemical hazards that may cause them harm. Workers should be trained to use chemical control measures properly, and how to check that the control measures are working. Workers should perform practice drills for cleaning up chemical spills safely. If workers need to use respirators, they need appropriate mask fitting and training.

个体防护装备(PPE)经常被用作(化学品)控制措施的一部分。应定期检查和维护个体防护装备(PPE)。损坏的个体防护装备不再能提供针对化学品危害的足够防护，应停止使用并加以维修或丢弃。

工作现场应确保在化学品储存区域备有适用的灭火器和(化学品)泄漏清理装备，并使用恰当的标识。

化学品的使用应从采购到过期加以追踪。过期化学品应加以识别并依据企业废物管理程序以安全的方式加以处置。这也应包括对过期化学品的包装材料进行安全处置。

- **应建立程序确保从事化学品相关工作的人员胜任**

工作现场应能证实其建立了相关程序。该程序确保设计、安装、维护和测试化学品控制措施的任何人员具备必要的技能、知识和经验，并且是胜任的。

工作人员应充分参与到化学品控制措施的开发过程中，确保他们有合适的装备来开展工作。应鼓励工作人员提出改进建议，并对他们发现的任何问题进行报告。

应给可能受到化学品伤害、需要了解化学品危害的工作人员和其他人员提供化学品危害信息及详尽指导。应对工作人员开展正确使用化学品控制措施以及如何检查控制措施是否有效的培训。工作人员应开展安全清除泄漏化学品的实战演练。如果工作人员需要使用呼吸器，他们需要合适的面罩和培训。

1.1.3 Related Life-Saving Rules

Life-Saving Rule 16: Do not smoke outside designated smoking areas

1.1.4 References

UK HSE, Working with substances hazardous to health A brief guide to COSHH, INDG136 (rev5), Revised Oct 2012.

1.2 Hydrogen Sulphide (Toxic Gas) Safety

1.2.1 Prompts

(1) Has a Safety Case been prepared inclusive of toxic gas risks, is it in place and operational?

(2) Have all toxic gas hazards been identified and risks assessed?

(3) Has a Quantitative Risk Assessment (QRA) for toxic gas been developed?

(4) Has a Toxic Gas Dispersion Study been performed?

(5) Have toxic gas bowties been developed?

(6) Are all sour service barriers and their performance standards in place and operational?

(7) Are sour service process containment barriers and systems functional and well maintained?

(8) Are toxic gas zoning and zoning controls for accidental releases and emergency response in place and operational?

1.1.3 相关保命法则

保命法则 16：禁止在指定区域外吸烟

1.1.4 参考资料

UK HSE，Working with substances hazardous to health A brief guide to COSHH，INDG136（rev5），Revised Oct 2012.

1.2 硫化氢(有毒气体)安全

1.2.1 督导提示

(1) 是否编制了包含有毒气体风险的安全例证？安全例证是否准备就绪并投入使用？

(2) 是否已识别所有有毒气体危害并评估了其风险？

(3) 是否针对有毒气体进行了定量风险分析(QRA)[1]？

(4) 是否进行了有毒气体扩散分析？

(5) 是否编制了有毒气体(相关事件)领结图？

(6) 所有含硫作业的屏障及其性能标准是否到位并投入使用？

(7) 含硫作业防泄漏类别的屏障和系统是否性能正常且维护良好？

(8) 针对意外泄漏和应急响应的有毒气体分区和分区控制措施是否到位并有效实施？

---

[1] QRA 也称作定量风险评估。

(9) Are toxic gas zoning and zoning controls for occupational health exposures in place and operational?

(10) Is a zonal toxic gas PTW system in place and operational?

(11) Are zonal toxic gas competency controls in place and operational?

(12) Is toxic gas awareness included as part of site induction training?

(13) Are personnel trained on the use and testing of personal H_2S detectors?

(14) Are personnel trained in the use of Self Contained Breathing Apparatus (SCBA)?

(15) Are emergency responders provided advanced training on H_2S search and rescue including the use of Respiratory Protective Equipment (RPE)?

(16) Are toxic gas detection barriers and systems functional and well maintained?

(17) Does toxic gas detection system cover all toxic gas hazard areas/buildings?

(18) Are detectors located in Heating, Ventilation and Air Conditioning (HVAC) inlets of all toxic gas threatened buildings?

(19) Has toxic gas detection mapping been carried out during design to optimise location of detectors?

(20) Do audible and visual alarms give warnings in areas of toxic gas accumulations?

(21) Do Cause and Effects drawings clearly define alarms and other functions that toxic gas detection initiates automatically or manually?

(9) 针对职业健康暴露的有毒气体分区和分区控制措施是否到位并有效实施？

(10) 分区有毒气体作业许可(PTW)制度是否到位并有效实施？

(11) 分区有毒气体(作业)能力控制措施是否到位并有效实施？

(12) 有毒气体意识培训是否作为其一部分包含在入场培训中？

(13) 作业人员是否接受过使用和测试个人硫化氢探测器的培训？

(14) 作业人员是否接受过使用正压呼吸器(SCBA)的培训？

(15) 是否为应急响应人员提供在硫化氢环境中进行搜救的高级培训？该培训是否包含使用呼吸防护装备(RPE)？

(16) 有毒气体探测类别的屏障和系统是否性能正常且维护良好？

(17) 有毒气体探测系统是否覆盖所有有毒气体危险区域/建筑物？

(18) 有毒气体探测器是否位于所有受到有毒气体威胁的建筑物的暖通空调(HVAC)入口处？

(19) 在设计过程中是否进行了有毒气体探测的定位分析以优化探测器的位置？

(20) 声光报警是否会在有毒气体聚集的区域发出警报？

(21) 因果图是否清楚地定义了有毒气体探测(器)自动或手动启动的警报及其他功能？

(22) Are calibration procedures clearly defined and how frequently are detectors re-calibrated? Is calibration frequency related to SIL?

(23) How are portable and personal toxic gas detectors controlled, tested and calibrated?

(24) Is information provided to explain how to raise and how to respond to the H_2S alarm?

(25) Is information provided about the colour and sound of an H_2S alarm?

(26) Are personnel required to wear personal H_2S monitors?

(27) Are sour service shutdown barriers and systems functional and well maintained?

(28) Are sour service protection barriers and systems functional and well maintained?

(29) Are sour service emergency response barriers and systems functional, well maintained and in an active state of readiness?

(30) Is there a wind sock visible from all areas of the site?

(31) Are there clearly marked escape and evacuation routes?

(32) Are emergency response drills carried out for H_2S releases?

(33) Are sour service lifesaving barriers and systems functional, well maintained and in an active state of readiness?

(34) Is H_2S emergency SCBA equipment readily and sufficiently available in the process area of the plant?

(35) Is H_2S emergency SCBA equipment readily and sufficiently available in the plant buildings?

（22）校准程序是否进行了明确定义？探测器重新校准的频率如何？校准频率是否与安全完整性等级（SIL）相关？

（23）如何控制、测试和校准便携式和个人有毒气体探测器？

（24）是否提供了信息解释如何发出和响应硫化氢警报？

（25）是否提供了有关硫化氢警报的颜色和声音信息？

（26）是否要求作业人员佩戴个人硫化氢探测器？

（27）含硫作业停车类别的屏障和系统是否性能正常且维护良好？

（28）含硫作业保护类别的屏障和系统是否性能正常且维护良好？

（29）含硫作业应急响应类别的屏障和系统是否性能正常且维护良好？是否处于准备就绪状态？

（30）现场所有区域是否都能看到风向标？

（31）是否有清晰标记的逃生和疏散路线？

（32）是否针对硫化氢泄漏开展应急演练？

（33）含硫作业救生类别的屏障和系统是否性能正常且维护良好？是否处于准备就绪状态？

（34）在场站的工艺装置区，为硫化氢紧急状况配置的正压呼吸器（SCBA）是否易于取用且数量充足？

（35）在场站的建筑物内，为硫化氢紧急状况配置的正压呼吸器（SCBA）是否易于取用且数量充足？

(36) Is a process to monitor, audit and improve toxic gas performance in place and operational?

1.2.2 Reading

Toxic gas within this topic refers to Hydrogen Sulphide (H_2S), a colourless gas commonly associated with oil and gas operations. H_2S is highly toxic and has been responsible for the deaths of many oil and gas workers worldwide, including many cases where people have died while attempting to rescue colleagues.

Fatality can occur after only a few seconds at higher concentrations but lower concentrations can still be fatal if the exposure time is increased. At concentrations of 500 to 1000 ppm, respiratory paralysis is preceded by a period of rapid breathing, and death will result unless the victim is removed from the contaminated area and given immediate medical treatment. Fatality occurs due to a lack of oxygen in the brain and heart tissues, through a mechanism similar to that of cyanide.

H_2S can be present in oil and gas process plants, pipelines, sweetening units, oil/gas/water wells (including during drilling or well workover), sewers, swamps or any area where there is decaying organic matter or stagnant water.

H_2S can also result from acid/caustic reactions, for example when acid is used to clean equipment containing iron sulphide or to neutralise spent caustic which has been used for H_2S scrubbing. H_2S may also be present in spent acid flowed back after well stimulation.

(36) 是否建立了监测、审核和改进有毒气体相关做法的程序并有效实施？

1.2.2 知识准备

本专题中的有毒气体是指硫化氢(H_2S)，一种通常与全球石油和天然气作业相关的无色气体。硫化氢具有高毒性，它一直是全世界石油和天然气行业许多死亡事件的原因，包括许多试图对同事进行救援(而发生死亡)的案例。

人在较高硫化氢浓度下仅在几秒钟后就会发生死亡，但如果暴露时间增加，较低的浓度仍会致命。当硫化氢的浓度为500~1000ppm($1ppm=10^{-6}$)时，人在呼吸麻痹之前会出现呼吸急促，除非将受害者从受污染区域移出并立即进行医疗处置，否则将导致死亡。致死原因是由于大脑和心脏组织缺氧，致死过程类似于氰化物中毒的机制。

硫化氢可存在于石油和天然气加工厂、管道、脱硫单元，油/气/水井(包括钻井或修井期间)、下水道、沼泽或任何有腐烂的有机物或积水的区域。

硫化氢也可以由酸/苛性反应产生，例如：当酸用于清洗含有硫化铁的设备或中和已用于洗涤硫化氢的废碱液时。硫化氢也可能出现在实施油气井增产措施之后回流的废酸液中。

Accidental release of H_2S associated with oil and gas facilities have the potential to lead to a Major Accident. A Major Accident can be defined as an "Uncontrolled Occurrence" which leads to severe or catastrophic effects to people, the environment, assets or the company's reputation.

The most significant opportunity to reduce H_2S risks exists during the design phase of the facility where hazards can be "designed out" or control systems can be put in place. During the operational phase, effective implementation of an HSE Management System (HSE - MS) and an Asset Integrity Management System (AIMS), supported by competent personnel in safety critical positions, are used to ensure that the risks associated with the operations of a facility are reduced to a level which can be considered As Low As Reasonably Practicable (ALARP).

Demonstrating to stakeholders (such as employees, shareholders, government entities, the public, etc.) that toxic gas release and FEX risks associated with hydrocarbon processing and storage are managed effectively is a highly complex process. Many International Oil Companies (IOCs) use the Safety Case approach which was developed under UK law. A Safety Case is a facility or operation-specific demonstration that risks (including H_2S risks) have been, or will be, reduced to "acceptable" or ALARP levels. It provides assurance that the company has sufficient systems (physical and procedural) in place to prevent, where practicable, major accidents from occurring, and where a major accident does occur, it has sufficient systems (again physical and procedural) and resources in place to effectively reduce the impact of that major accident on people, the environment, assets or the company's reputation.

与石油和天然气设施相关的硫化氢意外泄漏有可能导致重大事故。重大事故可被定义为对人员、环境、资产和企业声誉造成严重或灾难性影响的“非受控事件”。

设施的设计阶段，是降低硫化氢风险的最重要时机，危险可以通过设计手段予以清除或将控制系统设计到位。在运行阶段，在安全关键岗位胜任人员的支持下，有效实施 HSE 管理体系(HSE-MS)和资产完整性管理体系(AIMS)，用于确保与设施运行相关的风险降低至可认为合理可行尽可能低(ALARP)的水平。

向利益相关者(如员工、股东、政府部门、公众等)证明，与烃类加工和储存相关的有毒气体泄漏和火灾爆炸(FEX)风险得到有效管理，是一个非常复杂的过程。许多国际石油公司(IOCs)使用根据英国法律制定的安全例证方法。安全例证是对特定设施或特定作业的风险(包括硫化氢风险)已经或将会降低到“可接受”水平或“合理可行尽可能低(ALARP)”水平的一种展示。它提供保证：企业有足够的系统(物理的和程序的)防止(在可行的情况下)发生重大事故。在发生重大事故时，企业有足够的系统(物理的和程序的)和资源到位，以有效减少重大事故对人员、环境、资产和企业声誉的影响。

The following are the main requirements of the Safety Case as they pertain to H_2S risk management:

- **H_2S risks should be formally evaluated and assessed**

The Safety Case should demonstrate that H_2S risks have been comprehensively evaluated and assessed.

Hazard Identification (HAZID) studies are generally the start of the evaluation process. The HAZID is a systematic approach to identifying the different hazards on a facility and, using a risk matrix, determines which have Major Accident Hazard potential and therefore, need further evaluation. The output of the HAZID is a Hazards and Effects Register (or Hazard Datasheet) which summarises all H_2S risks and provides details of how they are being managed.

H_2S risks are evaluated as part of the Quantitative Risk Assessment (QRA) which is an integral part of the Safety Case documentation.

The Safety Case should demonstrate that H_2S hazards and risks have been comprehensively evaluated and assessed. Appropriate facility specific consideration should be given to the types of H_2S events which could occur, i. e. levels of concentration, time of exposure, etc.

- **H_2S bowties should be developed and operational**

Bowties provide a highly visual and graphical representation of H_2S risk management. Bowties show the hard barriers (i. e. equipment) and soft barriers (i. e. procedures) that are used to prevent, detect, mitigate, provide emergency response or have lifesaving function for Major Accidents involving H_2S. Bowties should be documented in the Safety Case.

当安全例证用于硫化氢风险管理时，以下是其主要要求：

- **应对硫化氢进行正式的危害评价和风险评估**

安全例证应展示已经对硫化氢危害进行了全面评价，对硫化氢风险进行了全面评估。

危害辨识(HAZID)分析一般是危害评价过程的开始。危害辨识(HAZID)是一种系统的方法，用于识别设施的不同危害，并使用风险矩阵确定哪些危害具有重大事故危害的可能性。重大事故危害需要进一步评估。危害辨识(HAZID)的输出是一个危害和影响登记表(或危害数据表)，该登记表(或数据表)汇总了所有硫化氢风险，并提供了如何管理这些风险的详细信息。

硫化氢风险作为定量风险分析的一部分进行评估，定量风险分析(QRA)是安全例证文档必不可少的组成部分。

安全例证应证实硫化氢危害已进行了全面评价并对其风险进行了全面评估。对可能发生的硫化氢事件类型(如浓度水平、暴露时间等)应结合设施具体情况进行恰当考虑。

- **应开发并应用硫化氢(相关事件)领结图**[1]

领结图为硫化氢风险管理提供了高度可视化和图形化的表现形式。领结图显示了用于预防和探测涉及硫化氢的重大事故、减轻其后果、提供应急响应或救生功能的硬屏障(如：设备)和软屏障(如：程序)。领结图应记录在安全例证中。

---

[1] 简称“硫化氢领结图”。

The use of bowties at the facility should be evaluated. This will require an assessment of the level of awareness and understanding of the facility specific bowties and to what degree they are operational.

Over time, barriers can degrade/deviate from performance standards or fail and become unavailable which can create risks which need to be identified, assessed, monitored and managed. For example, degradation of process containment barriers increases the risk of loss of containment events.

A fully operational bowtie should be routinely used as the basis for operations or integrity based risk assessments which can be threat line and barrier based. A fully operational bowtie should be integrated into the day to day work of site managers and technical staff with safety critical roles and responsibilities and be part of the operating culture. Barrier anomalies, defects and failures should automatically initiate operating bowtie reviews and risk assessments with the barrier status being updated accordingly.

- **H_2S hardware barriers (i. e. HSECES) should be identified and operational**

The hard barriers in the bowtie are known as HSE Critical Equipment and Systems (HSECES).

An HSECES is defined as any structure, plant, equipment, or system whereby their failure could cause or contribute substantially to a Major Accident, or whose purpose is to prevent or limit the effect of a Major Accident from the relevant events (i. e. a H_2S event).

应对领结图在生产设施中的使用情况进行评价。这将需要评估(工作人员)对于特定生产设施领结图的认识和理解水平以及领结图的应用程度。

随着时间的推移，屏障可能会低于/偏离其性能标准或者失效而不起作用，这会产生(新的)风险，这些风险需要辨识、评估、监控和管理。例如，工艺防泄漏类别的屏障性能降低，增加工艺泄漏事件的风险。

操作(任务)或(设施)完整性的风险评估，可基于威胁和屏障，其常规做法是把充分实施和运行的领结图作为其基础。充分实施和运行的领结图应被整合到承担安全关键角色和职责的现场管理人员和技术人员的日常工作中，并成为运营文化的一部分。屏障异常、缺陷和失效时，应自动对(设施)运行领结图进行审查和风险评估，并相应更新屏障状态。

• 硫化氢硬件屏障(即：HSE关键设备和系统)应加以确定且性能正常

领结图中的硬屏障被称为HSE关键设备和系统(HSECES)。

HSE关键设备和系统(HSECES)被定义为任何结构、设施、设备或系统，其失效可能引起或在很大程度上导致重大事故，或者其目的是预防或限制起因于相关事件(如：硫化氢相关事件)的重大事故的影响。

HSECES are normally grouped into different barrier types responsible for Structural Integrity, Process Containment, Ignition Control, Protection, Detection, Shutdown, Emergency Response and Lifesaving Systems. HSECES with a functional role in managing H_2S risks should be clearly shown on the H_2S bowtie.

Performance Standards provide parameters that are set and measured so that the suitability and effectiveness of HSECES can be assured and verified. They are essential requirements against which the HSECES must be maintained throughout the lifecycle of the installation.

An HSECES Register is developed based on the identification of HSECES with a functional role in managing H_2S risks in the bowties. Performance standards should be developed for each HSECES, with the register used to determine which performance standard applies to which HSECES.

The HSECES Register should be used to drive inspection, testing and maintenance activities and tasks at site. The objective of the inspection, testing, and maintenance activities is to ensure that HSECES maintain their functional specification and remain fit for purpose.

There should be a system in place to identify where HSECES are not available. Long term deviations or failures from Performance Standards that remain uncontrolled should be identified and recorded.

- **Structural Integrity controls should be in place and operational**

The objective of structural integrity HSECES is to prevent failure which could cause structural damage and/or instability of the installation leading to loss of containment.

HSE 关键设备和系统(HSECES)通常分为不同的屏障类型，负责结构完整性、工艺防泄漏、点火控制、保护、探测、停车、应急响应和救生系统(等功能)。硫化氢领结图应清楚地显示具有硫化氢风险管理功能的 HSE 关键设备和系统(HSECES)。

性能标准提供设置和测量的参数，用以确保和验证 HSE 关键设备和系统(HSECES)的适用性和有效性。它们是在生产设施的生命周期内维护 HSE 关键设备和系统(HSECES)必须遵照的基本要求。

基于领结图中识别的 HSE 关键设备和系统(HSECES)，建立具有硫化氢风险管理功能的 HSE 关键设备和系统(HSECES)登记册。每个 HSE 关键设备和系统(HSECES)都应建立性能标准，并在登记册中明确二者的关联。

HSE 关键设备和系统(HSECES)登记册应用于驱动现场检查、测试和维护活动及任务。检查、测试和维护活动的目标是，确保 HSE 关键设备和系统(HSECES)保持其功能规格并适合其目的。

确定 HSE 关键设备和系统(HSECES)在哪些地方不可用的制度应到位。HSE 关键设备和系统(HSECES)相对于其性能标准的、非受控的、长期偏离或失效，应予以识别和记录。

- **结构完整性类别的控制措施应到位且性能正常**

结构完整性类别的 HSE 关键设备和系统(HSECES)的目的是防止某些失效，这些失效可能引起结构破坏和/或装置不稳定，进而导致工艺泄漏。

Structural Integrity HSECES, typically associated with the processing and storage of hydrocarbons containing H_2S, include foundations and fixing equipment (such as pipe racks) which support any equipment which may contain hydrocarbons under normal or abnormal operations. All identified Structural Integrity HSECES should be in place and should be operational.

- **Process Containment controls should be in place and operational**

The objective of a Process Containment system is to prevent a release of hydrocarbons that could lead to a Major Accident.

Process Containment HSECES, typically associated with the processing and storage of hydrocarbons containing H_2S, include pressure vessels, heat exchangers, rotating equipment, piping, pipelines, tanks, closed drains and pressure relief systems. Process containment systems are the principle means for keeping process fluids inside the plant equipment and preventing loss of containment events that may lead to an H_2S release.

Internal corrosion of hydrocarbon containing systems can occur when H_2S and free water are present in the hydrocarbon fluid. In simple terms, the H_2S combines with the water creating acid like substances that attack the carbon steel. Inspection and monitoring of wall thicknesses (of pipes, vessels, etc.) can help predict the degree of corrosion and allow companies to implement corrosion management techniques to reduce corrosion and prevent release of hydrocarbons.

The HSECES for containment systems should include the necessary inspection intervals to ensure internal corrosion due to H_2S can be identified and managed.

通常与含有硫化氢的烃类加工和储存相关的结构完整性类别的 HSE 关键设备和系统(HSECES)，包括地基和固定其他设备的设施(例如管架)，支撑在正常或异常操作下可能含有烃类的任何设备。所有已确定的结构完整性类别的 HSE 关键设备和系统(HSECES)应到位且性能正常。

- **工艺防泄漏类别的控制措施应到位且性能正常**

工艺防泄漏系统的目标是防止可能导致重大事故的烃类泄漏。

通常与含有硫化氢的烃类加工和储存相关的工艺防泄漏类别的 HSE 关键设备和系统(HSECES)包括压力容器、热交换器、动设备、配管、管道、罐、闭式排放系统和压力释放系统❶。工艺防泄漏系统是将工艺流体保持在设备设施内并防止可能导致释放硫化氢的泄漏事件的主要手段。

当烃类流体中存在硫化氢和游离水时，含烃类系统可能发生内部腐蚀。简单来说，硫化氢与水结合，形成酸性物质，侵袭碳钢。检查和监测壁厚(管道、容器等)可以帮助预测腐蚀程度，有助于企业在此基础上实施腐蚀管理技术，减少腐蚀和防止烃类物质的泄漏。

用于防泄漏系统的 HSE 关键设备和系统(HSECES)应包括必要的检查间隔，以确保可以识别和管理由硫化氢引起的内部腐蚀。

---

❶压力释放系统也称之为减压系统。

All identified Process Containment HSECES should be in place and should be operational.

- **Site H_2S zoning and zoning controls should be in place and operational**

Facility and unit based zoning relies on the fact that the risk of fatality (or serious irreversible harm) reduces with distance from the location of an accidental toxic gas release from the process or wells.

Check to see whether the site has introduced H_2S zoning and access controls. A Red Zone should be defined if there is a possibility of H_2S release from process or wells which could cause a fatality after a relatively short exposure. A Yellow Zone should be defined if there is a possibility of H_2S release from process or wells which may be injurious to personnel after a relatively short exposure. H_2S zone boundaries are defined either by consequences of a release or the risk associated with those releases.

The toxic concentrations associated with the Red Zone have the possibility of causing death within a few minutes or leading to immediate fatalities. At this level "permanent in-place safety features" is required. Containment systems in Red Zones should be as far as practicable "leak free". Active controls systems should be preferred in Red Zones. Manual controls should be prohibited unless shown by risk assessment to be ALARP.

A Yellow Zone is considered to exist where persons affected would have time to don Personal Protective Equipment (PPE) carried ready for use, but would not have time to reach the nearest PPE location. The longer travel time for toxic concentrations to reach workers in the Yellow Zone allows for the possibility to rely more on controls that facilitate effective emergency response.

所有已确定的工艺防泄漏类别的HSE关键设备和系统(HSECES)应到位并运行良好。

• 作业场所硫化氢分区和分区控制措施应到位并有效实施

基于设施和单元的(硫化氢)分区有赖于以下事实：死亡风险(或严重的不可逆损害)随着人员跟地面设施或井下有毒气体意外释放位置之间的距离增加而减少。

检查作业场所是否已引入硫化氢分区和访问控制。如果有可能从地面设施或井下释放硫化氢，并可能会因为相对较短的暴露时间而导致死亡，则应定义为红色风险区域。如果有可能从地面设施或井下释放硫化氢，并可能会因为相对较短的暴露时间而对人员有害，则应定义为黄色风险区域。硫化氢分区边界由硫化氢泄漏的后果或与这些泄漏相关的风险来定义。

红色风险区域相关的毒性浓度有可能在几分钟内导致死亡或者导致立即死亡。在这些毒性浓度(风险级别)，应要求"永久性就位的安全措施"。只要可行，红色风险区域的防泄漏系统应"无泄漏"。红色风险区域应首选主动式控制系统。除非风险评估显示为合理可行尽可能低(ALARP)，否则应禁止手动控制。

在黄色风险区域，受影响的人有时间穿上随身携带的个体防护装备(PPE)，但没有时间到达最近的个体防护装备(PPE)存放位置。对黄色风险区域的工作人员来说，(需要)在有毒浓度中驻留的时间越长，就有可能越多地依赖促进有效应急响应的控制措施。

Occupational Health and Safety Zones are often considered for locations where workers are routinely exposed to levels of airborne chemicals (in gas, vapour or dust form). The design intent of the Process Containment systems in oil and gas processing facilities is to prevent a releases of hydrocarbons. As such, although it may be reasonable to discuss exposure to H_2S at concentrations which can have significant effects, long term exposure to H_2S at low concentrations should not be considered as long term releases of hydrocarbon gas will not be expected.

The use of Breathing Apparatus (BA) and personal H_2S detectors should not be the "only controls" to manage H_2S risks in operations.

A broad range of plant layout, simultaneous operations (SIMOPS), emergency response, competence controls and operating procedures must be considered in addition to BA and personal H_2S detectors. These controls must be effectively implemented (where shown to be ALARP).

- **A PTW system should be in place and operational**

As H_2S is combined with hydrocarbon gas within all of the facilities, the control of work with respect to H_2S can follow the overall PTW and JHA systems. The risk assessment should specifically identify H_2S as a hazard if it is expected to be present (in the same way it identifies hydrocarbon gas). Where H_2S is identified as a hazard, specific controls will need to be identified.

All isolations undertaken in Red and Yellow Zones must be controlled by permit.

在工作人员日常暴露于空气中不同浓度的悬浮化学物质(气体、蒸气或粉尘形式)的地方，经常要考虑职业健康和安全分区。在石油和天然气加工设施中，工艺防泄漏系统的设计意图是防止烃类泄漏。因此，讨论暴露于有显著影响的硫化氢浓度可能是合理的，但是不应考虑长期暴露于低浓度硫化氢，因为预期不会长期释放烃类气体。

使用呼吸器(BA)和个人硫化氢探测器，不应成为作业过程中管理硫化氢风险的唯一措施。

除呼吸器(BA)和个人硫化氢探测器外，还必须考虑各种不同的设施布局、同时作业(SIMOPS)[1]、应急响应、(人员)能力控制和操作程序。这些控制措施必须有效实施[达到合理可行尽可能低(ALARP)的风险水平]。

- **作业许可(PTW)制度应到位并有效实施**

由于在所有生产设施中硫化氢与烃类气体结合在一起，关于硫化氢相关工作的控制可以遵循统一的作业许可(PTW)和工作危害分析(JHA)制度。如果预计存在硫化氢，则风险评估应明确将硫化氢识别为(生产设施的)一种危险(与识别烃类气体的方式相同)。一旦硫化氢被识别为(生产设施的)一种危险，则应确定针对性的控制措施。

在红色风险区域和黄色风险区域进行的所有隔离必须通过作业许可(PTW)进行控制。

[1] SIMOPS 也称作“同步作业”或“交叉作业”。

A double block and bleed consisting of two separate block valves with or without a spectacle blind is recommended for all systems containing toxic fluids under pressure.

Isolation, depressurisation and purging of the main system followed by local isolation is always preferred. This reduces the number of isolations against high pressure or toxic material and the amount of valving needed.

- **H_2S competency controls should be in place and operational**

All personnel who carry out design, visit, work or carry out special (e. g. emergency) duties within Red and Yellow Zones must have a level of competence compatible with the tasks they are expected to carry out.

Competency controls require that H_2S related competencies be identified, training provided and that performance against requirements be monitored and reviewed at regular intervals.

- **Toxic gas detection systems should be in place and operational**

The objectives of toxic gas detection systems are to provide continuous automatic monitoring functions to alert personnel of the presence of H_2S and to allow control actions to be initiated, manually or automatically, in order to minimize the likelihood of escalation.

Detection HSECES, typically associated with the processing and storage of hydrocarbons containing H_2S, include toxic gas detection, process controls and alarms. All identified Detection HSECES should be in place and should be operational.

对所有含带压有毒流体的系统，推荐采用由两个单独的隔离阀组成的双阀导淋隔离，并根据情况加或不加八字盲板。

主系统隔离、泄压和吹扫后，优先采用局部隔离。这将减少针对高压或有毒物料的隔离次数以及所需阀门的数量。

- **硫化氢相关的能力控制程序应到位并有效实施**

在红色和黄色风险区域内进行设计、访问、工作或执行特殊（例如应急）任务的所有人员必须具备与他们预期执行任务相适应的能力水平。

硫化氢相关的能力控制程序要求：应确定与硫化氢相关的能力并提供培训；应定期监测和审查（能力）要求的落实情况。

- **有毒气体探测系统应到位且性能正常**

有毒气体探测系统的目的是提供连续的自动监测功能，当存在硫化氢时进行报警，并允许手动或自动启动控制动作，以便最小化事态升级的可能性。

通常与含硫化氢的烃类物质加工和存储相关的（有毒气体）探测类别的 HSE 关键设备和系统（HSECES），包括有毒气体探测、过程控制和报警。所有已确定的探测类别的 HSE 关键设备和系统（HSECES）都应到位且性能正常。

The specification of the toxic gas detection system should describe the basis for determining the location, number, and types of detectors. This requires a process of identifying and assessing the possible H_2S hazardous events in each area and evaluating the requirements to reliably detect these events.

The toxic gas detection devices should be selected taking into account their response characteristics and the conditions which can be experienced when detection is required. All toxic gas field devices should be suitable for the area in which they are located.

The toxic gas detection system should be designed, located, or protected so that it will be available in those emergencies where toxic gas detection is required.

Safety critical alarms and instruments should be provided with a Safety Integrity Level (SIL) rating.

Personal H_2S detectors should be available with mandatory use expected for all personnel at site. Awareness level training must be provided on the use and testing of personal H_2S detectors before being allowed access to site.

- **Toxic gas protection systems should be in place and operational**

The objective of toxic gas protection systems is to mitigate the effects of H_2S events.

Toxic gas protection HSECES, typically associated with the processing and storage of hydrocarbons containing H_2S, include chemical injection, flare systems, ventilation (forced), and Heating, Ventilation and Air Conditioning (HVAC) systems.

All identified toxic gas Protection HSECES should be in place and should be operational.

有毒气体探测系统的规范应描述确定探测器位置、数量和类型的基本原则。这需要一个过程用以识别和评估每个区域可能的硫化氢危险事件并评价可靠地探测这些事件的要求。

选择有毒气体探测装置时，应考虑其响应特性和需要探测时可能出现的情况。所有有毒气体现场设备应适合它们所在的区域。

对有毒气体探测系统应进行设计、定位或保护，以使其在需要探测有毒气体的紧急情况下可用。

应提供安全关键警报和仪表的安全完整性等级(SIL)水平。

在(有硫化氢危险的)作业场所，应为所有人员提供并强制要求佩戴个人硫化氢探测器。在允许访问(上述)作业场所之前，必须提供个人硫化氢探测器使用和测试的基本培训。

- **有毒气体保护系统应到位且性能正常**

有毒气体保护系统的目的是减轻硫化氢事件的影响。

通常与含硫化氢的烃类物质加工和存储相关的有毒气体保护类别的 HSE 关键设备和系统(HSECES)，包括化学药剂注入、火炬系统、强制通风、暖通空调(HVAC)系统。

所有已确定的有毒气体保护类别的 HSE 关键设备和系统(HSECES)应到位且性能正常。

- **Shutdown systems should be in place and operational**

The objectives of shutdown systems are to shutdown the process once specific conditions are reached (pressure, temperature, etc.) to prevent a loss of containment of H_2S or to minimise the extent of a loss of containment event should one occur and thereby, mitigate the effects of an H_2S release.

Shutdown HSECES, typically associated with the processing and storage of hydrocarbons containing H_2S, include emergency depressuring, High Integrity Pressure Protection System (HIPPS) and Process ESD Valves.

All identified Shutdown HSECES should be in place and should be operational.

- **Emergency escape and evacuation systems should be in place and in an active state of readiness**

The objective of emergency escape and evacuation systems is to allow personnel to readily escape or evacuate from a location following an H_2S release event.

Emergency escape and evacuation HSECES, associated with onshore hydrocarbon processing and storage facilities, typically include escape & evacuation routes, muster points, emergency lighting, communication system, uninterruptible power supply (UPS) and emergency power.

All identified emergency escape and evacuation HSECES should be in place and in an active state of readiness.

- **停车系统应到位且性能正常**

停车系统的目的是在达到特定条件(压力、温度等)时关停工艺流程,以防止硫化氢泄漏,或最大限度地减少泄漏事件发生时的损失程度,从而减轻硫化氢泄漏的影响。

通常与含硫化氢的烃类物质加工和存储相关的停车类别的HSE关键设备和系统(HSECES),包括紧急泄压(设备和系统)、高完整性压力保护系统(HIPPS)和工艺紧急切断(ESD)阀。

所有已确定的停车类别的HSE关键设备和系统(HSECES)应到位且性能正常。

- **应急逃生和撤离系统应到位并处于准备就绪状态**

应急逃生和撤离系统的目的是,在硫化氢泄漏事件发生后,允许人员快捷地从一个地点逃生和撤离。

与陆上烃类物质加工和存储设施相关的应急逃生和撤离类别的HSE关键设备和系统(HSECES),通常包括逃生和撤离路线、集合点、应急照明、通信系统、不间断电源(UPS)和应急供电。

所有已确定的应急逃生和撤离类别的HSE关键设备和系统(HSECES)应到位并处于准备就绪状态。

When an H_2S alarm sounds, it is an expected evacuation procedure to walk crosswind until out of the gas cloud and then turn upwind to a safe muster location. As the wind direction changes frequently it is important to have multiple egress routes from work locations and to test the escape plans on a frequent basis. Ensure that there is a wind sock visible from all areas of the site to allow identification of the wind direction.

Mustering following an H_2S emergency alarm allows for site management to account for all personnel in order to initiate rescue if people are missing. Mustering must be to safe (from toxic gas release) locations. Workers should be aware of those Muster Areas and should be well drilled in toxic emergency situations.

- **Lifesaving systems should be in place and in an active state of readiness**

The objective of a lifesaving system is to increase the likelihood of personnel to escape and evacuate the facility following a H_2S event.

Lifesaving HSECES, associated with onshore hydrocarbon processing and storage facilities, typically include Personal Survival Equipment (PSE), Lifesaving Medical Equipment and rescue vehicles.

All identified Lifesaving HSECES should be in place and should be in an active state of readiness.

The use of lifesaving equipment in Red and Yellow Zones should be identified and adequate numbers of equipment should be made available for all foreseeable emergency situations.

当硫化氢警报响起时，预期的撤离程序是(首先)垂直于风向移动直到有毒气体云团外，然后逆风移动到安全的集合地点。由于风向经常变化，重要的是工作地点要有多个出口路线并经常测试逃生计划。确保从作业现场的所有区域都能看到风向标，以便识别风向。

在硫化氢紧急警报之后，(紧急)集合允许现场负责人清点所有人员，以便在人员失踪时启动救援。(紧急)集合必须到安全的(远离有毒气体释放)位置。工作人员应该清楚那些紧急集合地点，并且应该在有毒(气体释放)的紧急情况下训练有素。

- **救生系统应位到并处于准备就绪状态**

救生系统的目的是增加人员在硫化氢事件发生后从生产设施逃生和撤离的可能性。

与陆上烃类物质加工和存储设施相关的救生类别的 HSE 关键设备和系统(HSECES)通常包括个人求生设备(PSE)、医疗救生设备和救援车辆。

所有确定的救生类别的 HSE 关键设备和系统(HSECES)都应到位并处于准备就绪状态。

应确定在红色风险区域和黄色风险区域中使用的救生设备，并应为所有可预见的紧急状况提供足够数量的救生设备。

- **Communication with local population should be undertaken to ensure adequate emergency response if it is required**

The American Industrial Hygiene Association (AIHA) provides guidance on the effects of H_2S on people. In relation to the public, the most stringent guidance should be used as they have no control over how the event is managed and have no training on what to do if H_2S is evident in the air.

The most stringent guidance is ERPG-1 (Emergency Response Planning Guide Level 1). This relates to the maximum airborne concentration below which nearly all individuals could be exposed for up to 1 hour without experiencing more than mild, transient adverse health effects or without perceiving a clearly defined objectionable odour. This level for H_2S is 0.1 ppm in air of H_2S.

Sites should identify scenarios where concentrations of H_2S in air could affect the public at concentrations above the ERPG-1 criteria. The type of information provided to the public, and the frequency of providing that information, should be identified for each site.

Short-term (also called acute) symptoms and effects are shown in Table 1-1.

• 应与当地居民进行沟通，以确保在需要时进行恰当的应急响应

美国工业卫生协会(AIHA)就硫化氢对人的影响提供了指南。对于公众而言，应该使用最严格的指南，因为他们无法控制事件如何管理，也没有接受过空气中明显出现硫化氢时该做些什么的培训。

最严格的指南是ERPG-1(紧急响应计划指南1级)。这与空气中的最大硫化氢浓度有关，低于该浓度，几乎所有个体都可以暴露长达1h，而不会经历比轻微的、短暂的不利健康影响更严重的情况，或没有察觉到明确的令人讨厌的气味。空气中硫化氢的上述最大浓度为0.1ppm(1ppm = 1×10^{-6})。

作业场所应确定空气中的硫化氢浓度在高于ERPG-1标准的浓度下可能影响公众的情景。每个作业场所应确定向公众提供的信息类型以及提供该类型信息的频率。

短期(也称作急性)症状和影响见表1-1。

Table 1-1 Short-term (also called acute) symptoms and effects

Concentration/ppm	Symptoms/Effects
0.00011-0.00033	Typical background concentrations
0.01-1.5	Odor threshold (when rotten egg smell is first noticeable to some). Odor becomes more offensive at 3-5 ppm. Above 30 ppm, odor described as sweet or sickeningly sweet.
2-5	Prolonged exposure may cause nausea, tearing of the eyes, headaches or loss of sleep. Airway problems (bronchial constriction) in some asthma patients.
20	Possible fatigue, loss of appetite, headache, irritability, poor memory, dizziness.
50-100	Slight conjunctivitis ("gas eye") and respiratory tract irritation after 1 hour. May cause digestive upset and loss of appetite.
100	Coughing, eye irritation, loss of smell after 2-15 minutes (olfactory fatigue). Altered breathing, drowsiness after 15-30 minutes. Throat irritation after 1 hour. Gradual increase in severity of symptoms over several hours. Death may occur after 48 hours.
100-150	Loss of smell (olfactory fatigue or paralysis).
200-300	Marked conjunctivitis and respiratory tract irritation after 1 hour. Pulmonary edema may occur from prolonged exposure.
500-700	Staggering, collapse in 5 minutes. Serious damage to the eyes in 30 minutes. Death after 30-60 minutes.
700-1000	Rapid unconsciousness, "knockdown" or immediate collapse within 1 to 2 breaths, breathing stops, death within minutes.
1000-2000	Nearly instant death

表 1-1 短期(也称作急性)症状和影响

浓度/ppm	症状/影响
0.00011~0.00033	典型的背景浓度
0.01~1.5	嗅觉阈值(臭鸡蛋气味刚能被闻出时的浓度)。在 3~5ppm 时气味最臭。超过 30ppm，气味为甜或甜腻味
2~5	长时间接触可导致恶心、流泪、头痛或失眠。可能诱发哮喘(支气管痉挛)
20	可能会出现疲劳、食欲不振、头痛、烦躁、记忆力差、头晕
50~100	1h 后出现轻度结膜炎(“气眼”)和呼吸道刺激症状。可能会导致恶心和食欲不振
100	咳嗽，眼睛刺激不适，2~15min 后出现嗅觉丧失(嗅觉疲劳)。15~30min 后出现不规则呼吸、嗜睡。1h 后出现喉咙肿痛等刺激症状。在随后的数小时内症状逐渐加重。可能会在 48h 后出现死亡
100~150	嗅觉丧失(嗅觉疲劳或瘫痪)
200~300	1h 后出现明显的结膜炎和呼吸道刺激症状。长时间暴露可能会导致肺水肿
500~700	站立不稳，5min 之内随即昏倒。30min 之内双眼受到严重损害。30~60min 后死亡
700~1000	迅速失去知觉，“摔倒”或者呼吸 1 下到 2 下后立即昏倒，呼吸骤停，在几分钟内死亡
1000~2000	几乎立刻死亡

注：1ppm = 1×10^{-6}。

1.2.3 Related Life-Saving Rules

Life-Saving Rule 7: Work with a valid work permit when required

Life-Saving Rule 8: Verify isolation before work begins and use the specified life protecting equipment

Life-Saving Rule 12: Conduct gas tests when required

1.2.4 References

(1) American Industrial Hygiene Association, AIHA Guideline Foundation, ERPG WEEL Handbook.

(2) https://www.osha.gov/SLTC/hydrogensulfide/hazards.html.

1.2.3 相关保命法则

保命法则 7：高危作业必须持有有效的作业许可

保命法则 8：开始作业之前确认隔离到位，并使用规定的防护装备

保命法则 12：按要求进行气体检测

1.2.4 参考资料

（1）American Industrial Hygiene Association，AIHA Guideline Foundation，ERPG WEEL Handbook.

（2）https：//www. osha. gov/SLTC/hydrogensulfide/hazards. html.

2

FIRE AND EXPLOSION PROTECTION

火灾爆炸保护

2.1 Fire Protection——Buildings

2.1.1 Prompts

(1) Have all fire hazards been identified, risks assessed and used to establish safe building?

(2) Is the fire, heat and smoke detection and warning system functional and well maintained?

(3) Are flame detectors (IR, UV/IR) available at designated locations?

(4) Are fire alarm systems regularly tested?

(5) Are fire drills held regularly and logged?

(6) Is the firefighting system functional and well maintained?

(7) Are all extinguishers fully charged and operable with latest and proper inspection tags?

(8) Are special purpose extinguishers clearly marked?

(9) Are all extinguishers and over-head sprinklers in good condition?

(10) Are fire monitors and hydrants unobstructed?

(11) Are fire monitors and hydrants free from leaks?

(12) Are the fire hose reels/fire hoses properly mounted and maintained in good operating condition?

(13) Are the fire hose reels/fire hoses tested regularly?

(14) Are nozzles in good condition?

(15) Are the correct number of nozzles and fire hoses available in fire hose boxes?

2.1 火灾保护——建筑物

2.1.1 督导提示

(1) 是否已识别所有火灾危险？是否已评估其风险并用于制定建筑消防安全措施？

(2) 火、热和烟雾探测及警报系统是否正常运行并得到良好维护？

(3) 火焰探测器（红外、紫外/红外）是否在指定位置可用？

(4) 火灾报警系统是否定期测试？

(5) 是否定期举行消防演习并加以记录？

(6) 灭火系统是否运行正常且维护良好？

(7) 所有灭火器是否正常充填并可用？是否有最新和正确的检查标签？

(8) 特殊用途灭火器是否有明显标记？

(9) 所有灭火器和在头顶上方的自动喷水灭火装置是否都处于良好状态？

(10) 消防水炮和消防栓是否畅通无阻？

(11) 消防水炮和消防栓是否有泄漏？

(12) 消防水带卷盘/消防水带是否正确安装并保持良好操作状态？

(13) 是否定期测试消防水带卷盘/消防水带？

(14) 消防水喷嘴是否状况良好？

(15) 消防水带箱内是否有正确数量的喷嘴和消防水带？

(16) Are fire main isolation valves maintained in good operating condition?

(17) Are water control valves and water pressure checked regularly?

(18) Are the fire water pumps operational and set on automatic mode?

(19) Are an adequate number of portable fire extinguishers available and easily accessible?

(20) Are sprinkler nozzle heads directed away from electrical switch boards and equipment?

(21) Is access to firefighting equipment unobstructed?

(22) Are emergency escape routes and exits functional and well maintained?

(23) Are fire doors and shutters in good operating condition?

(24) Are fire doors self-closing with latch in the closed position?

(25) Are fire doors unobstructed and protected against obstructions?

(26) Is exit signage properly illuminated at all locations?

(27) Is the emergency lighting functional and well maintained?

(28) Are materials stored away from a fire door opening (general recommendation is one metre)?

(29) Is a fire emergency evacuation plan in place?

(30) Has fire safety information been provided to employees and other people on the premises?

(31) Is staff fire safety training periodically provided?

(32) Is the fire risk assessment periodically reviewed and updated?

(16) 消防主隔离阀是否保持良好的运行状态？

(17) 是否定期检查消防水调节阀和消防水压？

(18) 消防水泵是否性能正常并设置为自动模式？

(19) 是否有足够数量的便携式灭火器可供使用且易于获取？

(20) 自动喷水灭火装置的喷嘴是否远离电气开关板和电气设备？

(21) 取用消防设备是否有障碍？

(22) 紧急逃生路线和应急出口是否可用并得到良好维护？

(23) 防火门和防火卷帘是否处于良好的运行状态？

(24) 防火门是否自行关闭？门闩是否在闭锁位置？

(25) 防火门是否畅通无阻并(有措施)防止其受阻？

(26) 所有场所的应急出口标志是否有适当照明？

(27) 应急照明是否运行正常且维护良好？

(28) 材料存储是否远离防火门开口(一般建议是 1m)？

(29) 火灾应急疏散计划是否到位？

(30) 是否向员工以及该场所内的其他人员提供了消防安全信息？

(31) 是否定期为工作人员提供消防安全培训？

(32) 是否定期审查和更新火灾风险评估？

2.1.2 Reading

Fire occurs as a result of a series of very rapid chemical reactions between a fuel and oxygen that releases heat and light. For combustion to occur, oxygen, an ignition source and a fuel source must all be present——commonly referred to as the "fire triangle". Flames and smoke are the visible manifestation of combustion.

Buildings need to be designed to offer an acceptable level of fire safety and minimise the risks from heat and smoke. The primary objective is to reduce the potential for fire related death or injury to the occupants of a building and others who may become involved, such as the fire and rescue service, to acceptable limits. The secondary objective is to protect contents and ensure that as much as possible, a building can continue to function after a fire and that it can be repaired. The risk to adjoining properties also needs to be considered.

- **A fire risk assessment should be performed for buildings at any site**

A fire risk assessment for buildings is an organised and systematic look at the site premises, the activities carried out and the likelihood that a fire could start and cause harm to people in and around the premises. Site management should ensure that site has carried out a fire safety risk assessment for all buildings and keeps it up to date.

Ensure that the fire risk assessment has identified all sources of fuel, ignition and oxygen (i.e. natural or mechanical ventilation).

2.1.2 知识准备

可燃物和氧气之间发生一系列非常快速的化学反应而释放出热和光，从而引发火灾。发生燃烧，必须存在氧气、点火源和可燃物——通常称为"火三角"。火焰和烟雾是燃烧的可见表现。

建筑物的设计需要达到可接受的消防安全水平，并将热量和烟雾产生的风险降至最低。设计的主要目标是将建筑物使用者和可能牵扯到的其他人员(例如消防和救援服务)发生火灾相关死亡或伤害的可能性降低到可接受的限度。设计的次要目标是保护建筑物并确保建筑物尽可能在火灾后继续发挥作用并且建筑物因火灾损坏的部分可以修复。设计还需要考虑火灾对相邻建筑物的风险。

- **任何场所内的建筑物都应进行火灾风险评估**

建筑物火灾风险评估是针对一个场所内的房屋、开展的活动、发生火灾的可能性以及火灾对房屋内和房屋周围人员造成伤害的可能性，所进行的有组织的、系统的调查。场所负责人应确保场所内所有建筑物都进行了火灾风险评估并使其保持最新状态。

确保火灾风险评估已识别了所有可燃物、点火源和氧气源(如：自然通风或机械通风)。

Ensure that the fire risk assessment has identified people in and around the premises and people especially at risk.

The fire risk assessment should identify a credible range of potential incidents from small fires in paper bins up to major fires that, though very unlikely, have the potential to place many people at risk. The fire risk assessment should evaluate radiative, convective and conductive modes of fire propagation and should also include the risks from smoke.

The fire risk assessment should fully investigate the effectiveness of measures to remove or reduce the hazard (i. e. the fuel, oxygen or ignition) and has assessed a broad range of fire protection measures.

- **A fire, heat and smoke detection and warning system should be in place and operational**

In larger premises, particularly those with more than one floor, where an alarm given from any single point is unlikely to be heard throughout the building, an electrical system incorporating sounders and manually operated call points (break-glass boxes) is likely to be required. This type of system is likely to be acceptable where all parts of the building are occupied; early fire detection and alarm raising would be made by building occupants.

确保火灾风险评估已识别了房屋内和房屋周围的人员，特别是处在危险当中的人员。

火灾风险评估应确定有可信度的一系列潜在事件，从废纸篓里的小火到重大火灾。后者虽然不太可能，却有可能将很多人置于危险之中。火灾风险评估应评价火灾蔓延的辐射、对流和传导模式，还应包括来自烟雾的风险。

火灾风险评估应充分调查消除或减少火灾危险(即可燃物、氧气源或点火源)的措施的有效性，并调查各种火灾保护措施。

- **火、热和烟雾探测及警报系统应到位且状态良好**

在较大的场所内，特别是那些有多个楼层的建筑物，任何单个点发出的警报不可能在整个建筑内都能听到，这时可能需要包含扬声器和手动报警按钮(需打破玻璃盖板的盒子❶)在内的电子系统。建筑物的所有部位都有居住者或使用者的时候，这种类型的电子系统可能是合适的，初期火灾探测和报警可以由居住者或使用者完成。

❶或者是需打开透明盖板的盒子。

Many types of fire, heat and smoke detection systems are available now. As with any system, there are advantages and disadvantages to all fire, heat and smoke detection systems. An appropriate and relevant system should be identified during building design that could combine multiple systems to ensure complete coverage of the entire building. Ensure back up power supply if the fire-detection and warning system is electrically powered.

- **Fire-fighting equipment should be in place and operational**

Firefighting equipment can reduce the risk of a small fire, such as a small fire in a waste-paper bin, developing into a large one. The safe use of an appropriate fire extinguisher to control a fire in its early stages can also significantly reduce the risk to people in the premises by allowing people to assist others who are at risk. The risk assessment should ensure that there are sufficient portable extinguishers, of the right type, for the risks identified in the building.

Ensure that the extinguishers are suitable for the purpose. Ensure that there are enough extinguishers sited throughout the premises at appropriate locations. Ensure that the right types of extinguishers are located close to the fire hazards. Ensure that users can get to them without exposing themselves to risk. Ensure that the extinguishers are visible or that their position is indicated using signage. Ensure that steps have been taken to prevent the misuse of extinguishers.

Some buildings could also have permanent installations of firefighting equipment such as hose reels for use by trained staff or firefighters.

有许多类型的火、热和烟雾探测系统可供选用。像任何系统一样，所有火、热和烟雾探测系统都有各自的优点和缺点。在建筑物设计期间，应为建筑物选择合适的系统，且可以包含许多不同类型的系统，以确保建筑物完全覆盖。如果火灾探测和警报系统是电动的，那么应确保有备用电源。

- **灭火设备应到位且状态良好**

灭火设备可以降低小火的风险，例如：(防止)废纸篓里的小火发展成大火。安全使用合适灭火器控制早期阶段火情，通过允许人们帮助处在危险中的其他人员，也可以显著降低房屋中人员的(火灾)风险。风险评估应确保有足够合适类型的便携式灭火器，用于建筑物中已识别的火灾风险。

确保灭火器满足要求。确保在整个房屋的适当位置有足够的灭火器。确保正确类型的灭火器位于火灾危险区域附近。确保灭火器使用者可以在不暴露于风险的情况下取用灭火器。确保灭火器可见或使用标牌指示灭火器的位置。确保已采取措施防止滥用灭火器。

一些建筑物还有永久安装的灭火设备，如(消防)水带卷盘，供训练有素的工作人员或消防员使用。

Other fixed fire-fighting equipment, such as fire fighter connections to fixed systems and fire suppression systems (such as sprinklers or gaseous systems using CO_2, FM200, etc.) may also be provided.

Ensure that a competent service engineer is used to inspect, test and maintain fire-fighting systems. Ensure that adequate testing, inspection and maintenance records have been kept.

Ensure that the necessary procedures are in place to maintain any fire-fighting facilities that have been provided for the safety of people in the building (or for the use of firefighters).

- **Escape routes and exits should be in place and operational**

Once a fire has started, been detected and a warning given, everyone in the premises should be able to escape to a place of total safety unaided by the fire and rescue service. However, people with disabilities and others with special needs may need help from staff who have been designated for the purpose.

Escape routes should be designed to ensure, as far as possible, that any person confronted by fire anywhere in the building, is able to turn away from it and escape to a safe location, e. g. a protected stairway. From there they will be able to go directly to a place of total safety away from the building.

还应提供其他固定式灭火设备，例如消防员与固定式消防系统的连接装置以及火灾抑制系统[例如自动喷水灭火装置或使用二氧化碳(CO_2)、七氟丙烷(FM200)等的气体灭火系统]。

确保使用胜任的服务工程师检查、测试和维护灭火系统。确保保留恰当的测试、检查和维护记录。

确保制定必要的程序，以维护为建筑物内的人员安全(或供消防员使用)而提供的灭火设施。

- **逃生路线和紧急出口应到位且状态良好**

一旦着火，(火情)被探测到并发出警报，建筑物内的每个人都应该能够逃离到一个(即便)没有火灾和救援服务的情况下(也是)完全安全的地方。然而，残疾人和其他有特殊需要的人可能需要专门指定人员的帮助。

逃生路线的设计应尽可能确保在建筑物内任何地方遇到火灾的任何人都应该能够远离火灾并逃到安全的地方，例如：受保护的楼梯。从那里，他们将能够直接前往远离建筑物的完全安全的地方。

The level of fire protection that should be given to escape routes will vary depending on the level of fire risk within the buildings and other related factors. Generally, buildings that are simple, consisting of a single storey, will require fairly simple measures to protect the escape routes, compared to a large multi-storey building, which would require a more complex and inter-related system of fire protection systems, such as fire zones segregated by fire doors.

People in the buildings must be able to find their way to a place of total safety if there is a fire by using escape routes that have adequate lighting. Emergency escape lighting is likely to be required in most areas of a building to combat the reduced visibility due to smoke.

Signs must be used, where necessary, to help people identify escape routes, find firefighting equipment and emergency telephones.

Ensure that there are regular checks on equipment provided to help maintain the escape routes.

- **A fire emergency evacuation plan should be in place and functional**

The purpose of an emergency plan is to ensure that the people in the building know what to do if there is a fire and that the buildings can be safely evacuated. An emergency plan should be based on the outcome of the fire risk assessment and should be available to all personnel in the building.

根据建筑物内火灾风险等级和其他相关因素，逃生路线的防火等级有所不同。通常，只有一个楼层构成的简单建筑物，需要相对简单的措施来保护逃生路线；而大型多层建筑则需要更复杂和相互关联的火灾保护系统，例如由防火门隔开的防火区。

如发生火灾，建筑物内的人员通过使用有足够照明的逃生路线，必须能够逃离到完全安全的地方。建筑物的大部分区域可能需要紧急逃生照明，以防止烟雾导致能见度降低。

必要时，必须使用消防标识帮助人们识别逃生路线、找到灭火设备和应急电话。

对提供用来帮助维护逃生路线的设备，确保定期进行检查。

- **火灾应急疏散计划应到位且有效**

应急(疏散)计划的目的是确保建筑物内的人员知道如果发生火灾该做什么，并且可以安全从建筑物内疏散。应急(疏散)计划应基于火灾风险评估的结果，并且建筑物内的所有人员都可得到。

- **Fire safety information should be provided to employees and other people at any site**

All staff should be provided with fire safety information and instruction as soon as possible after they are appointed. Ensure that staff who work outside of normal working hours, such as contract cleaners or maintenance staff are also included.

The information and instructions provided must be in a form that can be understood. The information provided should take account of those with disabilities such as hearing or sight impairment and be multi-lingual.

- **Fire safety training should be periodically provided**

Site management must ensure adequate fire safety training is provided for their staff.

The type of training provided should be customized to the specific and relevant features of the buildings and should:

① Address the findings of the fire risk assessment.

② Explain the emergency procedures.

③ Take account of the work activity and explain the duties and responsibilities of staff.

④ Also be provided as periodic refresher training as and where appropriate.

⑤ Be easy to understand for the staff and any other persons included in the training.

⑥ Be tested by fire drills.

- **Fire risk assessment should be periodically reviewed and updated**

The fire risk assessment should be periodically reviewed in order to determine how effectively the risk is being controlled.

- **任何场所应向其工作人员和(场所内的)其他人员提供消防安全信息**

所有工作人员被指派(到一个工作场所)后，应尽快向他们提供消防安全信息和指南，确保在正常工作时间以外工作的人员也包含在内，如来自承包商的清洁工或维修人员。

提供的消防安全信息和指南必须采用可以理解的形式。所提供的信息应考虑到那些有听力或视力障碍的残障人士，并应该是多语言版本。

- **应定期提供消防安全培训**

场所负责人必须确保为其工作人员提供恰当的消防安全培训。

提供的培训类型应基于建筑物详细的特征，并应：

① 考虑火灾风险评估的结果。

② 解释应急程序。

③ 考虑工作活动并解释工作人员的义务和责任。

④ 在适当的时候和适当的地方进行定期复训。

⑤ 工作人员和任何其他可能在场的人员易于理解。

⑥ 通过消防演习进行测试。

- **应定期审查和更新火灾风险评估**

应定期审查火灾风险评估，以确认火灾风险控制的有效程度。

If the fire risk assessment is no longer valid or there has been a significant change in the buildings that has affected fire precautions, a review and update of the fire risk assessment is required.

Reasons for review and update could include:

① Changes to work processes or the way they are organised, including the introduction of new equipment.

② Alterations to the building, including the internal layout.

③ Substantial changes to furniture and fixings.

④ The introduction, change of use or increase in the storage of hazardous substances.

⑤ The failure of fire protection.

⑥ A significant increase in the number of people present.

The potential risk of any significant change should be assessed before it is introduced. It is usually more effective to minimise a risk by design (such as ensuring adequate, appropriate storage space for an item before introducing it into the buildings).

2.1.3 Related Life-Saving Rules

Life-Saving Rule 16: Do not smoke outside designated smoking areas

如果火灾风险评估不再有效或建筑物发生重大变化影响到防火措施，则需要对火灾风险评估进行(重新)审查和更新。

审查和更新的原因可能包括：

① 工作流程或工作组织方式的变化，包括引进新设备。

② 对建筑物的改变，包括内部布局。

③ 家具和固定装置的重大变化。

④ 引入危险物质、改变危险物质的使用，增加危险物质的储存。

⑤ 火灾保护失效。

⑥ 建筑物内的人员数量大幅增加。

任何重大变更(变化)的潜在风险应在引入之前进行评估。通过设计手段最小化风险通常更为有效(例如：在将一件物品引入建筑物之前，确保该物品具有足够的、合适的存储空间)。

2.1.3 相关保命法则

保命法则 16：严禁在非吸烟区之外吸烟

2.2 Fire and Explosion Protection——Fuel Storage Tanks

2.2.1 Prompts

(1) Have all Fire and Explosion (FEX) hazards been identified and risks assessed?

(2) Are safety critical procedures and work instructions for FEX protection in place and operational?

(3) Are the fuel storage tanks and loading/unloading connections leak free?

(4) Is the fire and gas detection system functional and well maintained?

(5) Is the emergency shut down system functional and well maintained?

(6) Is the emergency relief system functional and well maintained?

(7) Is the Hazardous Area Classification (HAC) system in place and operational?

(8) Are the firefighting systems functional and well maintained?

(9) Are the bunding and drainage systems functional and well maintained?

(10) Are emergency response systems in place and in an active state of readiness?

(11) Is the tank loading and unloading system functional and well maintained?

2.2 火灾爆炸保护——燃油储罐

2.2.1 督导提示

(1) 是否已识别了所有火灾爆炸(FEX)危害并评估了其风险?

(2) 火灾爆炸(EEX)保护的安全关键程序和工作指南是否到位并有效实施?

(3) 燃油储罐和装/卸油连接线(含接头)是否无泄漏?

(4) 火灾和气体探测系统是否运行正常且维护良好?

(5) 紧急停车系统是否运行正常且维护良好?

(6) 紧急泄放系统是否运行正常且维护良好?

(7) 危险分区(HAC)制度是否到位并有效实施?

(8) 灭火系统是否运行正常且维护良好?

(9) 防泄漏围堰[1]和排水系统是否运行正常且维护良好?

(10) 应急响应系统是否到位并处于准备就绪状态?

(11) 储罐装卸系统是否运行正常且维护良好?

---

[1]在某些场合可以是“(防泄漏)栈板”或“(防泄漏)围堤”。

(12) Is the inspection and maintenance system in place and operational?

(13) Is there a process in place to ensure that people who work with fuel storage tanks are competent?

2. 2. 2 Reading

Fires and explosions (FEX) are likely to occur when flammable vapours or liquids are released from a controlled environment (a tank) to areas where there may be an ignition source, or, alternatively, when an ignition source is introduced into a controlled environment.

In this regard, fuel is considered to be refined hydrocarbons such as diesel, gasoline, kerosene, petroleum spirits, etc. and fuel storage tanks are considered to be fixed or mobile tanks with capacities over 1000 litres. Very large storage tanks within process plants are considered as part of those facilities and as such, are outside the scope of this topic.

Fuel storage tank facilities considered by this topic include:

① Above and below ground fixed bulk storage tanks with inventories of flammable fuels stored in individual tank or groups of tanks greater than 1000 litres.

② Portable or skid-mounted vessels with storage capacities greater than 1000 litres.

(12) 检查和维护制度是否到位并有效实施?

(13) 确保燃油储罐工作人员胜任的程序是否到位并有效实施?

2.2.2 知识准备

当易燃蒸气或液体从受控环境(储罐)释放到可能存在点火源的区域时，或者当点火源被引入到受控环境时，可能发生火灾爆炸(FEX)。

在本专题中，燃油被认为是精制烃，例如柴油、汽油、煤油、石油精等，燃油储罐被认为是容量超过1000L的固定或移动罐。油气处理设施内的超大型储罐被视为油气处理设施的一部分，不在本专题范围内。

本专题考虑的燃油储罐设施包括:

① 地上和地下固定式大容量储罐，储存在单个储罐或一组储罐中的燃油库存量超过1000L。

② 储存容量大于1000L的移动式或橇装式容器。

In general, within the oil and gas industry, fuel storage tanks, which contain fluids at atmospheric temperatures and pressures, are considered lower risk than processing facilities. The exception to this is where very large amounts of fuels are stored in dedicated fuel storage facilities with many tanks containing thousands, or even millions of barrels of liquid.

Fuel storage tanks are considered a lower risk due to the limited inventory and reduced ignition potential. However, fires and explosions can, and do, occur in these facilities and as such, adequate precautions and operational controls need to be in place.

The most significant opportunity to reduce FEX risks exists during the design phase of the facility where hazards can be "designed out" or control systems can be put in place. During the operational phase, effective implementation of an HSE Management System (HSE – MS) and an Asset Integrity Management System (AIMS), supported by competent personnel in safety critical positions, are used to ensure that the risks associated with the operations of a facility are reduced to a level which can be considered As Low As Reasonably Practicable (ALARP).

- **A process should be in place to identify FEX hazards and to assess risks**

The main FEX hazards associated with fuel tank storage facilities are the hydrocarbon fuels.

一般而言，在石油和天然气工业中，含有常温常压流体的燃油储罐被认为比烃类处理设施的风险低。例外情况是，非常大量的燃油储存在专用设施中，专用设施包含许多储罐，储罐中含有数千甚至数百万桶[1]燃油。

由于库存量有限和着火可能性减少，燃油储罐被认为风险较低。但是，这些设施可能并且确实会发生火灾和爆炸，因此需要采取适当预防措施和运行控制措施。

设施的设计阶段，是降低火灾爆炸(FEX)风险的最重要时机。危险可以通过设计手段予以清除或将控制系统设计到位。在运行阶段，在安全关键岗位胜任人员的支持下，有效实施 HSE 管理体系(HSE-MS)和资产完整性管理体系(AIMS)，用于确保与设施运行相关的风险降低到合理可行尽可能低(ALARP)的水平。

- **应建立程序辨识火灾爆炸(FEX)危害并评估其风险**

与燃油储存设施相关的主要火灾爆炸(FEX)危害是烃类燃料。

[1]这里的“桶”是一种计量单位。

A risk assessment should be carried out regardless of the quantity of flammable liquid present at the workplace. The risk assessment should include routine activities within the workplace as well as non-routine activities, e.g., maintenance work, that often has a greater potential for fire and explosion incidents. No further action is needed in cases of low risk or no risk to safety from fire and explosions.

The assessment should include risks arising from the tank contents, from the associated operational activities and risks to the tank from external sources. The likelihood of an accidental release of a FEX hazard is highest during transfer operations. Transfer operations relates to both the transfer from the bulk delivery system (such as a road tanker) and from the storage tank to the end user (such as vehicles and day tanks integrated in generators, etc.).

The aims of the risk assessment are to determine:

① How the flammable liquid could be released.

② The consequences of such a release.

③ How such a release could be ignited.

④ How to prevent the leak from happening.

⑤ How to mitigate the consequences of a release.

⑥ How to prevent ignition of the release.

⑦ Any external risks which may be present.

无论工作场所存在的易燃液体的数量如何，都应进行风险评估。除了评估工作场所内的日常活动外，风险评估还应涵盖非常规活动，例如维护工作，此类工作发生火灾和爆炸事件的可能性通常较高。如果没有火灾和爆炸方面的安全风险，或者风险微不足道，则无须采取进一步行动。

风险评估应包括来自储罐内容物和储罐相关作业活动的风险，以及储罐外部来源的风险。在转油作业期间，火灾爆炸(FEX)危害意外释放的可能性最高。转油作业涉及从散装运输系统(例如公路罐车)和从燃油储罐到最终用户(例如车辆以及发电机集成的油箱❶等)转移燃油。

风险评估的目的是确定：

① 易燃液体如何释放。

② 这一释放的后果。

③ 如何点燃这一释放。

④ 如何防止释放(溢出或泄漏)发生。

⑤ 如何减轻释放的后果。

⑥ 如何防止释放被点燃。

⑦ 可能存在的任何外部风险。

---

❶Day Tank 的意思是日常供应罐或日常供应油箱。

- **FEX control measures should be in place and operational**

The FEX risk assessment should consider the following control measures in relation to the fuel storage tank:

① Containment——Flammable liquids should be stored in tanks or containers constructed to a national or international standard to ensure their strength and integrity.

② Separation——Protects people and property from the effects of a fire at the tank, and protects the tank from fires which may occur elsewhere on site.

③ Ventilation——Ensures that any flammable vapours given off from a spill, leak or release will be rapidly dispersed.

④ Control of Ignition Sources——In certain areas, flammable atmospheres may occur either during normal operation or due to accidental spills or leakage. These areas are called hazardous areas and measures to control the introduction of sources of ignition are required.

⑤ Relief Vents——During normal tank operation, the pressure in the tank may vary. Pressures may increase during filling or if the ambient temperature rises. Conversely pressures may drop during emptying or with temperature falls. The tank venting system should provide: normal pressure relief, normal vacuum relief and emergency pressure relief.

⑥ Secondary Containment (Bunding)——The probability of a large leak from a well-designed and maintained fuel storage tank is low. However, the consequences of a spillage of flammable liquid are potentially high. Therefore, measures to contain spillages or leaks from storage tanks both during filling and other normal operations are essential.

• 火灾爆炸(FEX)控制措施应到位并有效实施

火灾爆炸(FEX)风险评估应考虑以下与燃油储罐相关的控制措施：

① 密闭——易燃液体应储存在罐或容器中，它们的构建应符合国家或国际标准，以确保其强度和完整性。

② 分隔——保护人员和财产免受燃油储罐火灾的影响，并保护燃油储罐免受现场其他地方可能发生火灾的威胁。

③ 通风——确保燃油溢出、泄漏或释放过程中产生的任何易燃蒸气将被迅速驱散。

④ 点火源控制——在某些区域，可能在正常操作期间或由于意外溢出或泄漏而产生易燃易爆环境。这些区域被称为危险区域，需要采取措施控制点火源的引入。

⑤ 释压通气——在油罐正常运行期间，油罐中的压力可能会有所变化。在充填期间或环境温度升高时，压力可能会增加。相反，在排空或温度下降时压力可能会下降。油罐通气系统应提供：正常压力释放❶、正常真空释放和紧急压力释放。

⑥ 次级防泄漏措施(围堤/围堰)——精心设计和维护的燃油储罐大量泄漏的可能性很低。然而，易燃液体溢出的后果可能很高。因此，在装油和其他正常操作期间，采取措施对储罐溢出或泄漏物进行控制是必要的。

---

❶Relief 根据上下文翻译为释放或泄放。

⑦ Detection Systems——Flammable gas detection should be considered as a means of providing early warning of a loss of containment.

⑧ Fire Protection——Active and passive fire protection measures can be provided by fire resistant claddings or coatings, fire walls, water cooling systems, foam blankets or extinguishing systems (including portable fire extinguishers for smaller tanks).

The FEX risk assessment should consider additional control measures in relation to the road tanker loading/unloading facility:

① To minimise the risk of overfilling, tankers should normally be loaded using a flow meter with a trip to stop the pump and close a shut-off valve automatically when a preset quantity has been delivered. Use of an independent high-level or overflow alarm is recommended to provide a warning if the meter fails. Meters may be protected by installing flow control valves.

② Before liquids are unloaded from tankers into storage tanks, particular attention should be paid to ensuring enough ullage is available in the storage tank to receive the load. The use of high-level alarms on storage tanks to avoid overfilling is generally recommended. To reduce the likelihood of spillage, the use of self-sealing couplings on the hose connections should be considered.

③ Precautions should be in place to prevent spillage due to vehicles being moved with the hoses still connected. This can be done by providing barriers such as wheel chocks, brake interlocks on the vehicle or breakaway couplings on the hose connections.

⑦ 探测系统——应将可燃气体探测视为提供泄漏事件早期预警的一种手段。

⑧ 火灾保护——可通过耐火覆层或涂层、防火墙、水冷系统、泡沫覆盖层或灭火系统(包括用于较小储罐的便携式灭火器)提供主动和被动火灾保护措施。

火灾爆炸(FEX)风险评估应考虑与油罐车装载/卸载设施相关的更多的控制措施:

① 为了尽量减少超量装填风险，通常应使用带跳闸(功能)的流量计来装载油罐车，以便在到达预设的燃油装填数量时自动停泵并关闭截止阀。建议使用独立的高液位或溢流警报，以便在流量计出现故障时发出警报。可以通过安装流量控制阀来保护流量计。

② 在将液体从油罐车卸载到储罐之前，应特别注意确保储罐中有足够的空余来接收油罐车所载燃油。通常建议在储罐上使用高液位警报以避免储罐超量装填。为减少溢出的可能性，应考虑在软管连接处使用自密封接头。

③ 应采取防范措施，防止在软管仍然连接时移动车辆造成(燃油)溢出。这可以通过提供屏障来实现，例如：轮楔[1]、油罐车上的制动联锁装置或软管接头上的断开式联轴节。

---

[1] 止轮块、轮制动垫块。

④ In addition to any automatic shut-off devices on the tanker or the tank, an emergency stop button should be installed at the loading/unloading point which will activate a quick-action shut-off valve or pump stop control. This will enable the operator/driver to immediately stop loading/unloading if a problem is observed.

⑤ Precautions against static electricity should be provided to prevent both ignition of spilled liquid should it occur and electric shock to operators and drivers. These include an earthing connection for the vehicle, electrically conducting hoses (to dissipate static buildup), elimination of splash filling, etc. For additional protection an interlock may be fitted to prevent operation of the control valve or loading pump until the earth connection is made. Splash filling may be avoided by ensuring that the fill pipe in a top-loading operation always remains submerged below the liquid level and also to the bottom of the tank or tanker.

⑥ Access gantries with stairs or ladders and protective handrails will usually be required for top-loading of vehicles. Gantries should be made of fire-resisting materials and, where necessary, should include means of access to the top of the tanker vehicles.

⑦ Fire protection measures should be designed to minimise the risk of fire spread. This may be achieved by ensuring that a ready supply of water and/or foam is available from fixed or mobile equipment. Hand extinguishers should also be located at appropriate points within the transfer area. The loading/unloading facilities should also be fitted with easily accessible fire alarm activation points.

④ 除了油罐车或油罐上的任何自动关断装置外，还应在燃油装载/卸载点安装紧急停止按钮，以启动快速截止阀或停泵控制。如果观察到问题，这将使操作人员/罐车驾驶人员能够立即停止装载/卸载燃油。

⑤ 应提供静电防范措施，以防止储罐发生溢出时溢出的燃油被点燃，以及防止对操作人员和罐车驾驶人员造成电击。静电防范措施包括车辆接地连接、导电软管(消除静电积聚)、消除燃油充填过程中的喷溅现象等。为了提供额外的保护，可以安装一个联锁装置，以防止控制阀或装油泵运行，直到完成接地连接。通过确保顶部装载操作过程中油管一直插到液面以下并到储罐或罐车的底部，可以避免燃油充填过程中的喷溅现象。

⑥ 车辆顶部装载燃油通常需要带有台阶或梯子以及防护扶手的龙门架。龙门架应由耐火材料制成，必要时应包括进入油罐车顶部的通道。

⑦ 应设计火灾保护措施用以尽量减少火势蔓延的风险。这可以通过确保从固定或移动式灭火设备中获得现成的水和/或泡沫来实现。手提式灭火器也应位于转油作业区域内的适当位置。装载/卸载设施还应配备易于触及的火警报警触发点。

- **Emergency response systems should be in place and in an active state of readiness**

The impact of an incident involving flammable liquids may be drastically reduced if prompt emergency action is taken. Everyone should know what to do in the event of spills, leaks or fires involving flammable liquids. Practical training and written procedures should be provided covering:

① Raising the alarm.

② Calling the fire and rescue service.

③ Controlling the spill or leak.

④ Tackling the fire (when it is safe to do so).

⑤ Evacuating the area safely.

Where an incident may affect people or property beyond the site boundary, interaction with the local emergency services may be required.

- **A process should be in place to control the inspection and maintenance of fuel tank storage facilities**

All facilities should be inspected for damage and deterioration. In addition to planned formal examinations, another inspection should be a combination of routine visual inspection by operators looking for obvious deterioration or damage.

There should be regular inspection and cleaning of interceptors, bunds, vents, slop tanks, loading and unloading facilities and any buildings where flammable vapour may be present. Where a defect is identified, there should be a system in place and operational to record this and to initiate appropriate remedial action.

- **应急响应系统应就位并处于准备就绪状态**

如果迅速采取应急措施，可能会大大减少涉及易燃液体的事件的影响。每个人都应该知道在发生易燃液体溢出、泄漏或火灾时该怎么做。应提供实用培训和书面程序，包括：

① 发出警报。

② 致电消防和救援服务。

③ 控制溢出或泄漏。

④ 控制火情(确保自身安全的情况下再去做)。

⑤ 安全撤离该区域。

如果事件可能影响工作场所边界以外的人员或财产，则可能需要联系当地应急服务部门。

- **应建立程序控制油罐存储设施的检查和维护**

应检查所有设施是否有损坏和性能降低❶。除了计划的正式检查外，另一种检查应该是由操作人员进行的、查找明显性能降低或损坏的常规目视综合检查。

应定期检查和清洁集油器、围堰、通气孔、污油罐、装卸设施以及可能存在易燃蒸气的建筑物。如果发现缺陷，应该有记录缺陷并启动适当补救措施的制度(可遵循)并有效实施。

---

❶Deterioration 也可翻译为劣化迹象。

Facilities and equipment shall be maintained in a safe condition. Storage tanks and all associated equipment, including walls and fences, should be properly maintained. Only personnel who are suitably qualified and authorised, and who fully understand the hazards, should carry out inspection and maintenance.

Safety critical equipment associated with the facilities should be listed on a preventive maintenance schedule, containing details of the scope and frequency of planned inspection and maintenance work. There should be periodic inspection of electrical equipment and the function testing of the isolation valves.

Firefighting equipment should be regularly maintained and, where appropriate, tested.

Examination of tanks, pipework and fittings should be carried out by a competent person.

A scheme of examination should be agreed between the user and the competent person, to include the scope and frequency of thorough examination. It is suggested that the intervals between internal examinations should be determined using a risk assessment approach based on tank service, maintenance history, known and potential damage mechanisms and their rates of attack.

Records should be kept of all examinations, tests, modifications and major maintenance. Schemes of examination should be in writing and should be reviewed regularly by the competent person and the equipment owner.

设施和设备应保持在安全状态。应妥善维护储罐和所有相关设备，包括墙壁和围栏。只有具备相应资质、得到相应授权且完全了解危险的人员才能进行检查和维护。

与设施相关的安全关键设备，应列在预防性维护计划中，其中包含与计划的检查和维护有关的工作范围及频次的详细信息。应定期检查电气设备并对隔离阀定期进行功能测试。

应定期维护消防设备，并在适当情况下进行测试。

罐体、管件和配件的检查应由胜任人员进行。

使用者与胜任(的检测)人员应商定检测计划，包括彻底检查的范围和频次。确定(储罐)内部检查间隔，建议使用基于储罐使用情况、维修历史、已知和潜在损伤机理及其损伤速率的风险评估方法。

应保留所有检查、测试、修改和重大维修的记录。检测计划应采用书面形式，并应由胜任(的检测)人员和设备负责人定期审查。

• A process should be in place to ensure that personnel who work with Fuel Storage Tanks are competent

A typical training programme should include the following:

① Hazards and properties of the liquids being stored.

② Safe operating procedures for the installation and its associated equipment.

③ The purpose of the safety features, including the importance of not removing or tampering with them.

④ The action to be taken if a fault in the equipment is detected

⑤ Dealing with minor leaks and spills.

⑥ The importance of good housekeeping and preventive maintenance.

⑦ Emergency procedures.

2.2.3 Related Life-Saving Rules

Life-Saving Rule 16: Do not smoke outside designated smoking areas

Life-Saving Rule 7: Work with a valid work permit when required

Life - Saving Rule 8: Verify isolation before work begins and use the specified life protecting equipment

Life-Saving Rule 9: Prevent dropped objects

Life-Saving Rule 12: Conduct gas tests when required

- **应建立确保燃油储罐工作人员胜任的程序**

典型的培训方案应包括以下内容：

① 储存液体的危害和特性。

② 设施及其相关设备的安全操作程序。

③ 安全装置的目的，包括不移除或不改动这些安全装置的重要性。

④ 检测到设备故障时应采取的措施。

⑤ 处理小的(燃油)泄漏和溢出。

⑥ 整洁和预防性维护的重要性。

⑦ 应急程序。

2.2.3 相关保命法则

保命法则 16：禁止在非吸烟区吸烟

保命法则 7：高危作业必须持有有效的作业许可

保命法则 8：开始作业之前确认隔离到位，并使用规定的防护装备

保命法则 9：防范坠落物

保命法则 12：按要求进行气体检测

2.2.4 References

(1) UK HSE, Storage of Flammable Liquids in Tanks, HSG176(Second edition)Published, 2015.

(2) UK HSE, Safety and Environmental Standards for Fuel Storage Sites, 2009.

2.3 Fire and Explosion Protection——Hydrocarbon Facilities

2.3.1 Prompts

(1) Has a Safety Case been prepared inclusive of FEX risks, is it in place and operational?

(2) Have all FEX hazards been identified and risks assessed?

(3) Has a Quantitative Risk Assessment(QRA)for FEX been developed, in place and operational?

(4) Does the QRA support operations and change management decision making at the facilities?

(5) Have FEX bowties been developed, in place and operational?

(6) Are all barriers and performance standards for FEX in place and operational?

(7) Has an HSECES Register been developed with Performance Standards?

(8) Is safety critical lifting equipment functional and well maintained?

2.2.4 参考资料

（1）UK HSE，Storage of Flammable Liquids in Tanks，HSG176(Second edition)Published，2015.

（2）UK HSE，Safety and Environmental Standards for Fuel Storage Sites，2009.

2.3 火灾爆炸保护——涉及烃类的生产设施

2.3.1 督导提示

（1）是否编制了包含火灾爆炸(FEX)风险的安全例证？安全例证是否准备就绪并投入使用？

（2）是否已识别所有火灾爆炸(FEX)危害并评估了其风险？

（3）是否针对火灾爆炸(FEX)进行了定量风险分析(QRA)？是否已完成？是否可用于实际操作？

（4）定量风险分析(QRA)是否支持设施运行和设施变更管理决策的制定？

（5）是否编制了火灾爆炸(FEX)领结图？是否已完成？是否可用于实际操作？

（6）火灾爆炸(FEX)的所有屏障和性能标准是否已就位？是否有效？

（7）是否编制了HSE关键设备和系统(HSECES)登记册？HSE关键设备和系统(HSECES)是否有配套性能标准？

（8）安全关键起重设备是否性能正常且维护良好？

(9) Are supporting and overhead structures maintained to a high degree of integrity?

(10) Have structural support failure mechanisms (through fatigue, vibration, abrasion or stress induced failure) been identified through inspection and monitoring?

(11) Is there adequate protection against physical impact to structural supports?

(12) Are FEX process containment barriers and systems functional and well maintained?

(13) Is Risk Based Inspection (RBI) used to prioritise where and how often corrosion inspection activities are performed?

(14) Do corrosion monitoring and protection systems cover all hydrocarbon containment systems?

(15) Are FEX ignition control barriers and systems functional and well maintained?

(16) Is the Hazardous Area Classification (HAC) system in place and operational?

(17) Are hazardous area zonal maps available and used as part of PTW system?

(18) Are sources of ignition effectively controlled in all hazardous areas by a combination of design, and systems of work?

(19) Are there adequate ignition controls in place for direct fired heaters, hot oil systems and processes operating above auto-ignition temperatures?

(20) Are fire and gas detection barriers and systems functional and well maintained?

(9) 支撑和架空结构是否保持了高度完整性?

(10) 是否通过检查和监测识别了结构支撑的失效机制(通过疲劳、振动、磨损或应力引起的失效)?

(11) 是否有足够的保护措施防止对结构支撑的物理影响?

(12) 防止火灾爆炸(FEX)的工艺防泄漏类别的屏障和系统是否性能正常且维护良好?

(13) 基于风险的检查(RBI)方法是否用来为腐蚀检查活动的执行地点和执行频率排列优先顺序?

(14) 腐蚀监测和保护系统是否覆盖所有烃类物料密闭系统?

(15) 防止火灾爆炸(FEX)的点火控制类别的屏障和系统是否性能正常且维护良好?

(16) 危险分区(HAC)制度是否到位并有效实施?

(17) 危险分区地图是否可获取并作为作业许可(PTW)制度的一部分使用?

(18) 通过设计措施和工作制度的组合,所有危险区域是否有效控制了点火源?

(19) 对于直燃加热器、热油系统和工作温度高于(工艺物料)自燃温度的工艺流程,是否有恰当的点火控制措施?

(20) 火灾和气体探测类别的屏障和系统是否性能正常且维护良好?

(21) Has flammable gas detection mapping been carried out to optimise location of detectors?

(22) Are flammable gas detectors provided in safe areas, including Control Room?

(23) Are fire and gas detectors functional, available and well maintained?

(24) Are fire and gas detector testing procedures and schedule clearly defined?

(25) Are fire and gas detector calibration procedures clearly defined and how frequently are detectors re-calibrated?

(26) Do Cause and Effects drawings clearly define alarms and other safety functions that fire and smoke detection initiate automatically or manually?

(27) Are fire and gas shutdown barriers and systems functional and well maintained?

(28) Has a recognised methodology (e. g. SIL, LOPA) been used to establish the required integrity of Emergency Shut Down (ESD) protective functions?

(29) Does the ESD system act separately and independently of the plant trip and alarm system?

(30) Are ESD protective functions subject to periodic testing (trip testing)?

(31) Is there a comprehensive testing procedure and schedule for ESD valves?

(32) Are full closure tests of all ESD valves performed?

(33) Are the FEX protection barriers and systems functional and well maintained?

（21）是否进行了可燃气体探测的定位分析以优化探测器的位置？

（22）控制室等安全区是否配置了可燃气体探测器？

（23）火灾和气体探测器是否性能正常、可用且维护良好？

（24）是否明确规定了火灾和气体探测器的测试程序和测试时间表？

（25）火灾和气体探测器的校准程序是否有明确定义？探测器重新校准的频率如何？

（26）“因果图”是否清楚地定义了由火灾和烟雾探测自动或手动启动的警报及其他安全功能？

（27）火灾和气体（泄漏）相关的停车类别的屏障和系统是否性能正常且维护良好？

（28）是否已采用公认方法确定所需紧急关停（ESD）保护功能的完整性要求？例如：安全完整性等级（SIL），保护层分析（LOPA）。

（29）紧急关停（ESD）系统是否独立于设施跳闸和报警系统单独起作用？

（30）紧急关停（ESD）保护功能是否需要定期测试（跳闸测试）？

（31）紧急关断（ESD）阀是否有综合测试程序和测试时间表？

（32）是否进行了所有紧急关断（ESD）阀的全封闭测试？

（33）火灾爆炸（FEX）保护类别的屏障和系统是否性能正常且维护良好？

(34) Is Passive Fire Protection (PFP) provided for selected structures in accordance with fire protection philosophy and fire hazard studies?

(35) Are critical valves, instrumentation and cables passively fire protected?

(36) Are procedures in place to ensure that both the passive fire protective system and the protected surface are regularly inspected and repaired as appropriate?

(37) Is all fire-fighting and active fire protection equipment available for immediate use in the event of fire?

(38) Are all fire-fighting and active fire protection equipment maintained in good working order?

(39) Where fixed fire protection systems are installed, is there an effective system of testing in compliance with code requirements?

(40) Are operating and maintenance procedures provided for fixed fire protection systems?

(41) Is the location of items such as the foam and water sources kept a safe distance from any hazardous installation?

(42) Is the fire water system supplied by a secure water supply which should include items such as backup diesel pumps where appropriate?

(43) Does the design ensure that the active fire protection system is not starved of water due to other demands on the water supply system during a fire?

(44) Have all fire-fighting equipment subject to inspection been suitably identified, e. g. numbering system?

（34）是否根据火灾保护理念❶和火灾危险分析为选定的构筑物提供被动火灾保护（PFP）？

（35）关键阀门、仪表和电缆是否有被动火灾保护？

（36）是否建立程序确保对被动火灾保护系统和受保护表面进行定期检查和酌情修理？

（37）所有灭火和主动火灾保护设备在发生火灾时是否可立即使用？

（38）所有灭火和主动火灾保护设备是否保持良好的工作状态？

（39）如果安装了固定式火灾保护系统，是否有符合规范要求的有效测试制度？

（40）是否为固定式火灾保护系统提供了操作和维护程序？

（41）泡沫和水源等灭火材料的位置与任何危险设施是否保持了安全距离？

（42）消防水系统是否由安全的供水系统供应？在适当时，安全的供水系统应包括备用柴油泵等项目。

（43）设计是否确保主动火灾保护系统不会由于着火时对供水系统的其他需求而缺水？

（44）是否对所有应检查的消防设备进行了适当标识？例如：编号系统。

---

❶火灾保护理念也称消防理念。

(45) Are fire pump tests run weekly? Are records kept and faults attended to immediately?

(46) Are FEX emergency response barriers and systems in place and in an active state of readiness?

(47) Is emergency lighting in Hazardous Areas certified for the zone classification?

(48) Are SCBA readily available in adequate numbers for use in emergency situations?

(49) Are FEX lifesaving barriers and systems in place and in an active state of readiness?

(50) Is a process to monitor, audit and improve FEX protection performance in place and operational?

2.3.2 Reading

Fires and explosions (FEX) caused by accidental releases of hydrocarbons associated with oil and gas facilities have the potential to lead to Major Accidents. A Major Accident can be defined as an "Uncontrolled Occurrence" which leads to severe or catastrophic effects to people, the environment, assets or the company's reputation.

(45) 是否每周进行一次消防泵测试？是否保留记录并立即处理故障？

(46) 火灾爆炸(FEX)应急响应类别的屏障和系统是否到位并处于准备就绪状态？

(47) 危险区域的应急照明是否经过危险分区认证？

(48) 正压呼吸器(SCBA)在紧急情况下是否足够和容易获取？

(49) 火灾爆炸(FEX)救生类别的屏障和系统是否到位并处于准备就绪状态？

(50) 是否建立了监测、审核和改进火灾爆炸(FEX)保护做法的程序并有效实施？

2.3.2 知识准备

石油和天然气设施相关的烃类物质意外释放引发的火灾爆炸(FEX)，有可能导致重大事故。重大事故可被定义为对人员、环境、资产或企业声誉造成严重或灾难性影响的“非受控事件”。

The most significant opportunity to reduce FEX risks exists during the design phase of the facility where hazards can be "designed out" or control systems can be put in place. During the operational phase, effective implementation of an HSE Management System (HSE-MS) and an Asset Integrity Management System (AIMS), supported by competent personnel in safety critical positions, are used to ensure that the risks associated with the operations of a facility are reduced to a level which can be considered As Low As Reasonably Practicable (ALARP).

Demonstrating to stakeholders (such as employees, shareholders, government entities, the public, etc.) that FEX risks, associated with hydrocarbon processing and storage, are managed effectively, is a highly complex process. Many International Oil Companies (IOCs) use the Safety Case approach which was developed under UK law. A Safety Case is a facility or operation-specific demonstration that risks (including FEX risks) have been, or will be, reduced to "acceptable" or ALARP levels. It provides assurance that the company has sufficient systems (physical and procedural) in place to prevent, where practicable, major accidents from occurring, and where a major accident does occur, it has sufficient systems (again physical and procedural) and resources in place to effectively reduce the impact of that major accident on people, the environment, assets or the company's reputation.

The following are the main requirements of the Safety Case as it pertains to FEX risk management:

设施的设计阶段，是降低火灾爆炸(FEX)风险的最重要时机，危险可以通过设计手段予以清除或将控制系统设计到位。在运行阶段，在安全关键岗位胜任人员的支持下，有效实施HSE管理体系(HSE-MS)和资产完整性管理体系(AIMS)，用于确保与设施运行相关的风险降低至可认为合理可行尽可能低(ALARP)的水平。

向利益相关者(如员工、股东、政府部门、公众等)证明，与烃类加工和储存相关的火灾爆炸(FEX)风险得到有效管理，是一个非常复杂的过程。许多国际石油公司(IOCs)使用根据英国法律制定的安全例证方法。安全例证是对特定设施或特定作业的风险(包括火灾爆炸(FEX)风险)已经或将会降低到“可接受”水平或“合理可行尽可能低(ALARP)”水平的一种展示。它提供保证：企业有足够的系统(物理的和程序的)防止(在可行的情况下)发生重大事故。在发生重大事故时，企业有足够的系统(物理的和程序的)和资源到位，以有效减少重大事故对人员、环境、资产和企业声誉的影响。

当安全例证用于火灾爆炸(FEX)风险管理时，以下是其主要要求：

• **FEX hazards and risks should be formally evaluated and assessed**

Hazard Identification (HAZID) studies are generally the start of the evaluation process. The HAZID is a systematic approach to identifying the different hazards on a facility and, using a risk matrix, determines which have Major Accident Hazard potential and therefore, need further evaluation. The output of the HAZID is a Hazards and Effects Register which summarises all FEX risks and provides details of how they are being managed.

The main FEX hazards associated with hydrocarbon facilities are the hydrocarbons themselves (oil and gas).

FEX risks are evaluated as part of the Quantitative Risk Assessment (QRA) which is an integral part of the Safety Case documentation.

The Safety Case should demonstrate that FEX hazards and risks have been comprehensively evaluated and assessed. Appropriate facility specific consideration should be given to the types of FEX events which could occur, i. e. flash fires, pool fires, jet fires, fireballs, vapour cloud explosions, tank fires, tank rim fires, etc.

• **FEX bowties should be developed and be operational**

Bowties provide a highly visual and graphical representation of FEX risk management. Bowties show the hard barriers (i. e. equipment) and soft barriers (i. e. procedures) that are used to prevent, detect, mitigate, provide emergency response or have a lifesaving function for Major Accidents involving FEX. Bowties should be documented in the Safety Case.

- **应对火灾爆炸(FEX)进行正式的危害评价和风险评估**

危害辨识(HAZID)分析一般是危害评价过程的开始。危害辨识(HAZID)是一种系统的方法，用于识别设施的不同危害，并使用风险矩阵确定哪些危害具有重大事故危害的可能性。重大事故危害需要进一步评估。危害辨识(HAZID)的输出是一个危害和影响登记表(或危害数据表)，该登记表(或数据表)汇总了所有火灾爆炸(FEX)风险，并提供了如何管理这些风险的详细信息。

与烃类设施相关的主要火灾爆炸(FEX)危害是烃类本身(石油和天然气)。

火灾爆炸(FEX)风险作为定量风险分析(QRA)的一部分进行评价，定量风险分析(QRA)是安全例证文档必不可少的组成部分。

安全例证应证明火灾爆炸(FEX)危害和风险已得到全面评价和评估。(安全例证)应恰当结合设施具体情况，考虑可能发生的火灾爆炸(FEX)事件类型，即闪火、池火、喷射火、火球、蒸气云爆炸、油罐着火，油罐边缘着火等。

- **应开发并应用火灾爆炸(事件)领结图**

领结图为火灾爆炸(FEX)风险管理提供了高度可视化和图形化的表现形式。领结图显示了用于预防和探测涉及火灾爆炸(FEX)的重大事故、减轻其后果、提供应急响应和救生功能的硬屏障(如：设备)和软屏障(如：程序)。领结图应记录在安全例证中。

The use of bowties at the facility should be evaluated. This will require an assessment of the level of awareness and understanding of the facility specific bowties and to what degree they are operational.

Over time, barriers can degrade/deviate from performance standards or fail and become unavailable which can create risks which need to be identified, assessed, monitored and managed. For example, degradation of process containment barriers increases the risk of loss of containment events.

A fully operational bowtie should be routinely used as the basis for operations or integrity based risk assessments which can be threat line or barrier based. A fully operational bowtie should be integrated into the day to day work of site managers and technical staff with safety critical roles and responsibilities and will be part of the operating culture. Barrier anomalies, defects and failures should automatically initiate operating bowtie reviews and risk assessments with the barrier status being updated accordingly.

- **FEX hardware barriers (i. e. HSECES) and their Performance Standards should be identified and be operational**

The hard barriers in the bowtie are known as HSE Critical Equipment and Systems (HSECES).

An HSECES is defined as any structure, plant, equipment, or system the failure of which could cause or contribute substantially to a Major Accident, or whose purpose is to prevent or limit the effect of a Major Accident(i. e. FEX) from the relevant events.

应对领结图在生产设施中的使用情况进行评价。这将需要评估(工作人员)对于特定生产设施领结图的认识和理解水平以及领结图的应用程度。

随着时间的推移，屏障可能会降低/偏离性能标准或失效而不起作用，这会产生(新的)风险，这些风险需要辨识、评估、监控和管理。例如：工艺防泄漏屏障性能降低，增加工艺泄漏事件的风险。

操作(任务)或(设施)完整性的风险评估，可基于威胁和屏障，其常规做法是把充分实施和运行的领结图作为其基础。充分实施和运行的领结图应被整合到承担安全关键角色和职责的现场管理人员和技术人员的日常工作中，并成为运营文化的一部分。屏障异常、缺陷和失效时，应自动对(设施)运行领结图进行审查和风险评估，并相应更新屏障状态。

• 火灾爆炸(FEX)硬件屏障(即 HSE 关键设备和系统)及其性能标准应加以确定且性能正常

领结图中的硬屏障被称为 HSE 关键设备和系统(HSECES)。

HSE 关键设备和系统(HSECES)被定义为任何结构、设施、设备或系统，其失效可能引起或在很大程度上导致重大事故，或者其目的是预防或限制起因于相关事件的重大事故[如：火灾爆炸(FEX)]的影响。

HSECES are normally grouped into different barrier types responsible for Structural Integrity, Process Containment, Ignition Control, Protection, Detection, Shutdown, Emergency Response and Lifesaving Systems. HSECES with a functional role in managing FEX risks should be clearly shown on the bowtie.

Performance Standards are parameters that are set and measured so that the suitability and effectiveness of HSECES can be assured and verified. They are essential requirements against which the HSECES must be maintained throughout the lifecycle of the installation.

An HSECES Register is developed based on the identification of HSECES in the bowties. Performance standards should be developed for each HSECES, with the register used to determine which performance standard applies to which HSECES.

The HSECES Register should be used to drive inspection, testing and maintenance activities and tasks at site. The objective of the inspection, testing, and maintenance activities is to ensure that the HSECES maintain their functional specification and remain fit for purpose.

There should be a system in place to identify where HSECES are not available. Long term deviations or failures from Performance Standards that remain uncontrolled should be identified and recorded.

- **Structural Integrity controls are in place and are operational**

The objective of structural integrity HSECES is to prevent failure which could cause structural damage and/or instability of the installation leading to loss of containment.

HSE关键设备和系统(HSECES)通常分为不同的屏障类型，负责结构完整性、工艺防泄漏、点火控制、保护、探测、停车、应急响应和救生系统(等功能)。火灾爆炸(FEX)领结图应清楚地显示具有火灾爆炸(FEX)风险管理功能的HSE关键设备和系统(HSECES)。

性能标准提供设置和测量的参数，用以确保和验证HSE关键设备和系统(HSECES)的适用性和有效性。它们是在生产设施的生命周期内维护HSE关键设备和系统(HSECES)必须遵照的基本要求。

基于领结图中识别的HSE关键设备和系统(HSECES)，建立HSE关键设备和系统(HSECES)登记册。每个HSE关键设备和系统(HSECES)都应建立性能标准，并在登记册中明确二者的关系。

HSE关键设备和系统(HSECES)登记册应用于驱动现场检查、测试和维护活动及任务。检查、测试和维护活动的目标是，确保HSE关键设备和系统(HSECES)保持其功能规格并适用其目的。

确定HSE关键设备和系统(HSECES)在哪些地方不可用的制度应到位。HSE关键设备和系统(HSECES)相对于其性能标准的、非受控的、长期偏离或失效，应予以识别和记录。

- **结构完整性类别的控制措施应到位且性能正常**

结构完整性类别的HSE关键设备和系统(HSECES)的目的是防止某些失效，这些失效可能引起结构损坏和/或装置不稳定进而导致工艺泄漏。

Structural Integrity HSECES, typically associated with hydrocarbon processing and storage, include foundations and fixed equipment (such as pipe racks) which support any equipment which may contain hydrocarbons under normal or abnormal operations. All identified Structural Integrity HSECES should be in place and should be operational.

- **Process Containment controls are in place and are operational**

The objective of a Process Containment system is to prevent a release of hydrocarbons that could lead to a Major Accident.

Process Containment HSECES, typically associated with hydrocarbon processing and storage facilities, include pressure vessels, heat exchangers, rotating equipment, pipelines, tanks, closed drains and relief systems. Process containment systems are the principle means for keeping process hydrocarbon fluids inside the plant equipment and preventing loss of containment events that may lead to FEXs. All identified Process Containment HSECESs should be in place and should be operational.

- **Ignition and site access controls should be in place and are operational**

The objective of the control of ignition is to minimize the likelihood of ignition of flammable liquids and gases following a loss of containment.

通常与烃类加工和储存相关的结构完整性类别的 HSE 关键设备和系统(HSECES)包括地基和固定其他设备的设施(例如管架)，支撑在正常或异常操作下可能含有烃类的任何设备。所有已确定的结构完整性类别的 HSE 关键设备和系统(HSECES)应到位且性能良好。

- **工艺防泄漏类别的措施应到位且性能正常**

工艺防泄漏系统的目标是防止可能导致重大事故的烃类泄漏。

通常与烃类加工和储存设施相关的工艺防泄漏类别的 HSE 关键设备和系统(HSECES)包括压力容器、热交换器、动设备、管道、罐、闭式排放系统和压力释放系统。工艺防泄漏系统是将工艺流程中的烃类流体保持在设施设备内并防止可能导致火灾爆炸(FEXs)泄漏事件的主要手段。所有已确定的工艺防泄漏类别的 HSE 关键设备和系统(HSECESs)应到位且性能正常。

- **点火和场站访问控制措施应到位并有效实施**

点火控制的目的是，尽量减少泄漏事件发生后易燃液体和气体被点燃的可能性。

Ignition Control HSECESs, typically associated with hydrocarbon processing and storage facilities, include hazardous area ventilation, non-hazardous area ventilation, certified electrical equipment, earth bonding, purge gas systems and miscellaneous ignition control components. Flares are generally not included as ignition sources as they should be designed as safe by location, however, some flare ignition systems may be considered as ignition sources (such as flame front ignition sources). All identified Ignition Control HSECES should be in place and should be operational.

The objective of Hazardous Area Classification (HAC) zoning is to avoid the ignition of those releases which inevitably occur from time to time in the operation of facilities handling flammable liquids and vapours. Classification of hazardous areas affects the type of electrical equipment permitted and hence can reduce sources of ignition. At the design stage, measures usually aim at minimising the range of the Hazardous Area Classification (HAC) zones. Consideration should also be given to hydrocarbon releases from non-routine operational activities and from major releases. Operations activities should ensure that these measures are not compromised by introducing materials (e.g. solvents, chemicals, etc.) that could give off a flammable vapour or by introducing activities that could provide an uncontrolled source of ignition.

Access to any area of a facility where hydrocarbon vapours could be present should be controlled by a central Permit to Work system.

通常与烃类物料加工和储存设施相关的点火控制类别的HSE关键设备和系统(HSECESs)包括危险区域通风、非危险区域通风、经认证的电气设备、接地连接、气体系统吹扫以及各种点火控制做法❶。火炬通常不作为点火源，因为根据其所在位置它们的设计应该是安全的。然而，一些火炬点火系统可能被视为点火源(例如地面传焰点火装置)。所有已确定的点火控制类别的HSE关键设备和系统(HSECES)应到位且性能正常。

危险分区(HAC)的目的是，避免点燃那些在易燃液体和蒸气处理设施运行过程中不时发生的释放。危险区域的分类影响允许的电气设备类型，从而可以减少点火源。在设计阶段，(点火)控制措施通常旨在最小化危险分区(HAC)的区域范围。还应考虑非常规活动和重大泄漏所引起的烃类物料释放。在引入可能释放易燃蒸气的材料(例如溶剂、化学品等)或引入可能包含非受控点火源的活动时，作业活动应确保遵守(点火)控制措施。

进入任何可能存在烃类蒸气的设施区域，应由统一的作业许可制度进行控制。

---

❶点火控制需要系统考虑，因此英文原文使用了component这个单词，意思是“系统的组成部分”。

Where vehicles regularly travel through the facilities, their potential as ignition sources should be assessed. Internal combustion engines (including the catalytic converters on the exhaust system) can generates sparks and hot surfaces with sufficient energy to ignite hydrocarbon vapours. Access control measures coupled with vehicle specification (e. g. diesel engines which provide hot surfaces only) reduce the likelihood of a vapour accumulation being ignited.

Access control for personnel should ensure that items capable of igniting flammable vapours are not permitted in production areas of the facilities. These items include matches, lighters, radios, cameras, mobile phones, etc.

Arrangements to minimise the likelihood of ignition should be provided in accordance with the requirements of relevant studies (QRA, HAC, etc.). This should include minimisation of the number of potential ignition sources as far as reasonable.

Ignition of flammable liquid and gas leaks should be minimised, providing equipment which is designed in accordance with a recognized standard.

Procedures to control the use of temporary electrical equipment which can present ignition sources should be established.

Direct fired equipment should be located or protected to prevent ignition following loss of containment.

- **Fire and gas detection systems should be in place and are operational**

The objectives of the Fire and Gas (F&G) detection systems are to:

如果车辆经常在生产设施间穿行，则应评估车辆作为点火源的可能性。内燃机(包括排气系统上的催化转化器)可以产生具有足够能量的火花和热表面，以点燃烃类物料的蒸气。与车辆规格相结合的进入控制措施(例如：仅提供热表面的柴油发动机)降低了烃类物料蒸气积聚被点燃的可能性。

确保人员出入生产设施不携带能够点燃易燃蒸气的物品。这些物品包括火柴、打火机、(非防爆)对讲机、(非防爆)相机、(非防爆)手机等。

应根据相关研究[定量风险分析(QRA)，危险分区(HAC)等]结论的要求，做出最小化点火可能性的安排。这应包括尽可能合理地减少潜在点火源的数量。

应提供设计符合公认标准的设备，以尽可能减少点燃泄漏的易燃液体和气体(的可能性)。

应建立使用可能产生点火源的临时电气设备的控制程序。

直接燃烧设备的位置或保护措施，应(能够)防止泄漏事件后点燃(泄漏的易燃液体和气体)。

- **火灾和气体探测系统应到位且性能正常**

火灾和气体(F&G)探测系统的目的是：

① Provide continuous automatic monitoring functions to alert personnel of the presence of a hydrocarbon fire or a flammable gas condition.

② Allow control actions to be initiated manually or automatically in order to minimize the likelihood of escalation.

Detection HSECES, typically associated with hydrocarbon processing and storage facilities, include fire and gas detection and process controls and alarms. All identified Detection HSECES should be in place and should be operational.

The location, number, and types of detectors should be determined by a formal study (such as Fire and Gas Mapping). This requires a process of identifying and assessing the possible F&G hazardous events in each area and evaluating the requirements to reliably detect these events.

The F&G detection devices should be selected taking into account their response characteristics and the conditions which can be experienced when detection is required. Fire detectors should be selected to be suitable for detection of the types of fires that can occur in the area. All F&G detection field devices should be suitable for the area in which they are located and should be approved for use by an external recognized authority.

The F&G detection control system should be designed, located, or protected so that it will be available in those emergencies where fire and gas detection is required.

Safety critical alarms and instruments should be provided with a Safety Integrity Level (SIL) rating.

① 提供连续的自动监测功能，以提醒工作人员烃类火灾或易燃气体的存在状况。

② 允许手动或自动启动控制措施，以尽量减少(事件)升级的可能性。

通常与烃类物料处理和储存设施相关的探测类别的 HSE 关键设备和系统(HSECES)包括火灾和气体探测以及过程控制和报警。所有已确定的火灾和气体探测类别的 HSE 关键设备和系统(HSECES)应到位且性能正常。

探测器的位置、数量和类型应通过正式分析(如火灾和气体探测定位分析❶)确定。这需要一个识别和评估每个区域可能的火灾和气体(F&G)危险事件的过程，并评估可靠探测这些事件的要求。

应根据其响应特性和需要探测时可能遇到的状况来选择火灾和气体(F&G)探测装置。应选择合适的火灾探测器，以便探测该区域可能发生的火灾类型。所有火灾和气体(F&G)探测现场设备，应适用于它们所在的区域，其使用应经过外部公认权威机构的批准。

应设计、定位或保护火灾和气体(F&G)探测控制系统，以便在需要进行火灾和气体(F&G)探测的紧急情况下可用。

安全关键警报和仪表应提供安全完整性等级(SIL)水平。

---

❶Fire and Gas Mapping 也可翻译为“火灾和气体探测器定位分析”或者“火灾和气体探测点定位分析”。

• **Fire protection systems should be in place and are operational**

The objective of fire protection systems is to mitigate the effects of FEX events.

Fire protection HSECES, typically associated with hydrocarbon processing and storage facilities, include deluge systems, firewater pumps, firewater ring mains, firewater tank, foam systems, sprinkler systems, gaseous fire suppression systems, passive fire protection, flare systems, and Heating, Ventilation, and Air Conditioning (HVAC) systems. All identified Fire Protection HSECES should be in place and should be operational.

• **Emergency shutdown systems should be in place and are operational**

The objectives of shutdown systems are to prevent loss of containment following process upset events and to mitigate the effects of an FEX.

Shutdown HSECES, typically associated with hydrocarbon processing and storage facilities, include emergency depressuring, High Integrity Pressure Protection System (HIPPS) and Process ESD Valves. All identified Shutdown HSECESs should be in place and should be operational.

- **火灾保护系统应到位且性能正常**

火灾保护系统的目的是减轻火灾爆炸(FEX)事件的影响。

通常与烃类加工和储存设施相关的火灾保护类别的 HSE 关键设备和系统(HSECES)包括雨淋系统、消防水泵、主消防环管、消防水罐、泡沫系统、自动喷水灭火系统、气体灭火系统、被动火灾保护系统、火炬系统、以及暖通空调(HVAC)系统。所有已确定的火灾保护类别的 HSE 关键设备和系统(HSECES)应到位且性能正常。

- **紧急停车系统应到位且性能正常**

停车系统的目的是防止工艺过程的不稳定事件升级为工艺泄漏事件并减轻火灾爆炸(FEX)的影响。

通常与烃类物料处理和存储设施相关的停车类别的 HSE 关键设备和系统(HSECES)包括紧急泄压(设备和系统)、高完整性压力保护系统(HIPPS)和工艺紧急切断(ESD)阀。所有已确定的停车类别的 HSE 关键设备和系统(HSECES)应到位且性能正常。

• **Emergency escape and evacuation systems should be in place and in an active state of readiness**

The objective of emergency escape and evacuation systems is to allow personnel to readily escape and evacuate from a location following an FEX event.

Emergency escape and evacuation HSECES, typically associated with hydrocarbon processing and storage facilities, include Temporary Refuge/Muster Areas, escape & evacuation routes, emergency lighting, emergency communication system, Uninterruptible Power Supply (UPS) and emergency power. All identified Emergency HSECES should be in place and in an active state of readiness。

• **Lifesaving systems should be in place and in an active state of readiness**

The objective of a lifesaving system is to increase the likelihood of personnel to escape and evacuate the facility following an FEX event.

Lifesaving HSECES, associated with onshore hydrocarbon processing and storage facilities, typically include Personal Survival Equipment (PSE), Lifesaving Medical Equipment and rescue vehicles. All identified Lifesaving HSECES should be in place and should be in an active state of readiness.

- **应急逃生和撤离系统应到位并处于准备就绪状态**

应急逃生和撤离系统的目的是，在火灾爆炸(FEX)事件发生后，允许人员快捷地从事发地点逃生和撤离。

通常与烃类物料处理和储存设施相关的应急逃生和撤离类别的 HSE 关键设备和系统(HSECES)包括临时避难所/紧急集合区、逃生和撤离路线、应急照明、应急通信系统、不间断电源(UPS)和应急供电。所有已确定的应急逃生和撤离类别的 HSE 关键设备和系统(HSECES)应到位并处于准备就绪状态。

- **救生系统应到位并处于准备就绪状态**

救生系统的目的是增加人员在火灾爆炸(FEX)事件发生后逃离和撤离设施的可能性。

与陆上烃类物料加工和储存设施相关的救生类别的 HSE 关键设备和系统(HSECES)通常包括个人求生设备(PSE)、救生医疗设备和救援车辆。所有已确定的救生类别的 HSE 关键设备和系统(HSECES)应到位并处于准备就绪状态。

2.3.3 Related Life-Saving Rules

Life-Saving Rule 16: Do not smoke outside designated smoking areas

Life-Saving Rule 7: Work with a valid work permit when required

Life-Saving Rule 8: Verify isolation before work begins and use the specified life protecting equipment

Life-Saving Rule 9: Prevent dropped objects

Life-Saving Rule 12: Conduct gas tests when required

2.3.3 相关保命法则

保命法则 16：禁止在非吸烟区吸烟

保命法则 7：高危作业必须持有有效的作业许可

保命法则 8：开始作业之前确认隔离到位，并使用规定的防护装备

保命法则 9：防范坠落物

保命法则 12：按要求进行气体检测

APPENDIX

附录

Appendix 1 Leadership Site Visit Prompt Card Sample

LEADER NAME <Insert Leader Name >
SITE NAME <Insert Location>
DATE <Insert Date >

LSV PROMPT CARD

<INSERT TOPIC>

LSV-PC-XX

PROMPT <Insert Prompt>	**SCORE**	**PROMPT** <Insert Prompt>	**SCORE**
Supporting Evidence?		***Supporting Evidence?***	
PROMPT <Insert Prompt>	**SCORE**	**PROMPT** <Insert Prompt>	**SCORE**
Supporting Evidence?		***Supporting Evidence?***	
PROMPT <Insert Prompt>	**SCORE**	**PROMPT** <Insert Prompt>	**SCORE**
Supporting Evidence?		***Supporting Evidence?***	
PROMPT <Insert Prompt>	**SCORE**	**PROMPT** <Insert Prompt>	**SCORE**
Supporting Evidence?		***Supporting Evidence?***	

Remark: For how to use LSV PROMPT CARD, please refer to relevant sections of the Book—Recommended Leadership Site Visit Program and Safety Models.

附录1 管理人员现场安全督导提示卡样例

管理人员姓名<填写管理人员姓名> **地点**<填写地点> **时间**<填写年月日>	**督导提示卡** <嵌入主题> LSV-PC-XX
督导提示 <嵌入督导提示> **打分**	**督导提示** <嵌入督导提示> **打分**
支持性证据?	***支持性证据?***
督导提示 <嵌入督导提示> **打分**	**督导提示** <嵌入督导提示> **打分**
支持性证据?	***支持性证据?***
督导提示 <嵌入督导提示> **打分**	**督导提示** <嵌入督导提示> **打分**
支持性证据?	***支持性证据?***
督导提示 <嵌入督导提示> **打分**	**督导提示** <嵌入督导提示> **打分**
支持性证据?	***支持性证据?***

注：管理人员现场安全督导提示卡的使用说明参见分册《安全督导推荐做法及常用安全模型》的有关章节。

SITE NAME

<Insert Location>

LSV PROMPT CARD

<INSERT TOPIC>

LSV-PC-XX

SAFETY CONVERSATION

As part of the site visit you will have safety conversations with the line organisation.
Use this form to summarise your observations.

Who did you talk to? (Provide position not name)
How was your safety conversation received?
How well did they understand the topic?
Were they able to show you procedures they use to control their work?
How well do they understand and implement the procedures?
What information/suggestions did you provide to them?
What improvements did they suggest?

NOTES

LEADER NAME	POSITION
SIGNATURE	DATE

地点

<填写地点>

督导提示卡

<嵌入主题>

LSV-PC-XX

安全交谈

作为现场安全督导工作的一部分，您将与现场人员进行安全交谈。
请使用这张表汇总您的交谈所得。

您与谁进行了交谈？（提供岗位而不是姓名）
你们是如何开展安全交谈的？
他们对本次交谈的主题理解得怎么样？
他们能否向您展示用于控制作业（危害）的程序？
他们对程序理解得怎么样？实施得怎么样？
您向他们提供了什么信息/建议？
他们提出了什么改进建议？

备注

管理人员姓名	岗位
签名	日期

Appendix 2 Abbreviations and Acronyms

附录 2 缩略语

AIChE	American Institute of Chemical Engineers
AIHA	American Industrial Hygiene Association
AIMS	Asset Integrity Management System
ALARP	As Low as Reasonably Practicable
BA	Breathing Apparatus
COSHH	Control of Substances Hazardous to Health
ERPG-1	Emergency Response Planning Guide Level 1
ESD	Emergency Shut Down
F&G	Fire and Gas
FEX	Fire and Explosion
H_2S	Hydrogen Sulphide
HAC	Hazardous Area Classification

HAZID	Hazard Identification
HIPPS	High Integrity Pressure Protection System
HSE	Health, Safety and Environment
HSECES	HSE Critical Equipment and Systems
HSE-MS	Health, Safety and Environment Management System
HVAC	Heating, Ventilation and Air Conditioning
IADC	International Association of Drilling Contractors
INDG	Industry Guidance
IOCs	International Oil Companies
IOGP	International Association of Oil & Gas Producers
JHA	Job Hazard Analysis
LOPA	Layer of Protection Analysis
LSVP	Leadership Site Visit Program
SDS	Safety Data Sheet for Chemical Products
OSHA	Occupational Safety and Health Administration

PFP	Passive Fire Protection
PPE	Personal Protective Equipment
ppm	Parts Per Million
PSE	Personal Survival Equipment
PTW	Permit to Work
QRA	Quantitative Risk Assessment
RBI	Risk Based Inspection
RPE	Respiratory Protective Equipment
SCBA	Self-Contained Breathing Apparatus
SIL	Safety Integrity Level
SIMOPS	Simultaneous Operations
UK HSE	UK Health and Safety Executive
UPS	Uninterruptible Power Supply

Appendix 3 English-Chinese Vocabulary

附录3 中英文对照词汇表

Active State of Readiness	准备就绪状态
American Industrial Hygiene Association(AIHA)	美国工业卫生协会
American Institute of Chemical Engineers	美国化学工程师协会
As Low as Reasonably Practicable (ALARP)	合理可行尽可能低
Asset Integrity Management System (AIMS)	资产完整性管理体系
Block Valve	隔离阀
Breakaway Coupling	断开式联轴节
Breathing Apparatus(BA)	呼吸器
Closed Drain	闭排(闭式排放)
Control of Substances Hazardous to Health(COSHH)	健康公害性物质控制
Designed Out	通过设计手段予以清除
Double Block and Bleed	双阀导淋隔离

Emergency Escape and Evacuation	应急逃生和撤离
Emergency Pressure Relief	紧急压力释放
Emergency Relief System	紧急泄放系统
Emergency Response Planning Guide Level 1(ERPG-1)	紧急响应计划指南1级
Emergency Shut Down (ESD)	紧急切断，紧急关停
Emergency Stop Button	紧急停止按钮
Escape and Evacuation Route	逃生和疏散路线
Fire and Explosion (FEX)	火灾爆炸
Fire and Gas (F&G)	火灾和气体
Fire and Gas Detection System	火灾和气体探测系统
Fire Hose	消防水带
Fire Propagation	火灾蔓延
Flame Front Ignition Source	地面传焰点火装置
Flare System	火炬系统
Hazard Identification (HAZID)	危害辨识
Hazardous Area Classification (HAC)	危险分区
Health, Safety and Environment	健康安全与环境
Health, Safety and Environment Management System (HSE-MS)	健康安全环境管理体系

Heating, Ventilation and Air Conditioning (HVAC)	暖通空调
High Integrity Pressure Protection System (HIPPS)	高完整性压力保护系统
Hose Reel	水带卷盘，软管卷盘
HSE Critical Equipment and Systems (HSECES)	HSE 关键设备和系统
Hydrogen Sulphide (H_2S)	硫化氢
Industry Guidance(INDG)	行业指引
Interceptor	集油器
International Association of Drilling Contractors	国际钻井承包商协会
International Association of Oil & Gas Producers	国际油气生产商协会
International Oil Companies (IOCs)	国际石油公司
Job Hazard Analysis (JHA)	工作危害分析
Layer of Protection Analysis (LOPA)	保护层分析
Leadership Site Visit Program	管理人员现场安全督导方案
Local Isolation	局部隔离
Loss of Containment Event	工艺泄漏事件

Flow Control Valve	流量控制阀
Major Accident	重大事故
Manually Operated Call Points	手动报警按钮
Safety Data Sheet for Chemical Products (SDS)	化学品安全技术说明书
Occupational Safety and Health Administration	(美国)职业安全与卫生管理局
Operating Bowtie	(设施)运行领结图
Overfilling	超量装填
Parts Per Million (PPM)	百万分之一
Passive Fire Protection (PFP)	被动火灾保护
Performance Standard	性能标准
Permanent In-place Safety Feature	永久就位的安全措施
Permit to Work (PTW)	作业许可
Personal Protective Equipment (PPE)	个体防护装备
Personal Survival Equipment (PSE)	个人求生设备
Pipework	管件
Pressure Relief System	压力释放系统、减压系统
Process Containment	工艺防泄漏
Process Containment Barrier	工艺防泄漏屏障

Process Containment Control	工艺防泄漏措施
Quantitative Risk Assessment (QRA)	定量风险分析、定量风险评估
Red Zone	红色风险区域
Relief Vents	释压通气
Risk Based Inspection (RBI)	基于风险的检查
Respiratory Protective Equipment (RPE)	呼吸防护装备
Safety Case	安全例证
Safety Feature	安全装置
Safety Integrity Level (SIL)	安全完整性等级
Scheme of Examination	检测计划
Self-Contained Breathing Apparatus (SCBA)	正压呼吸器
Self-Sealing Couplings	自密封接头
Shutdown System	停车系统
Shut-off Device	关断装置
Shut-off Valve	截止阀
Simultaneous Operations (SIMOPS)	同时作业，同步作业，交叉作业
Site Induction Training	入场培训

Spectacle Blind	八字盲板
Sprinkler	(建筑物内的)自动喷水灭火装置
Structural Support	结构支撑
Systems of Work	工作制度
Thorough Examination	彻底检查
Toxic Gas Dispersion Study	有毒气体扩散分析
Toxic Gas Zoning	有毒气体分区
Trip	跳闸
UK Health and Safety Executive	英国健康安全执行局
Uncontrolled Occurrence	非受控事件
Uninterruptible Power Supply (UPS)	不间断电源
Well Stimulation	增产措施
Wheel Chock	轮楔、止轮块、轮制动垫块
Wind Sock	风向标
Yellow Zone	黄色风险区域
Zoning Controls	分区控制措施